D1604665

Heavy Minerals in Colour

To the memory of Professor Janet Watson
and Emile

Heavy Minerals in Colour

Maria A. Mange

Department of Earth Sciences
University of Oxford

Heinz F. W. Maurer

Hofstattweg 11
CH-3044 Säriswil
Switzerland

CHAPMAN & HALL
London · New York · Tokyo · Melbourne · Madras

Published by Chapman & Hall, 2–6 Boundary Row, London SE1 8HN

Chapman & Hall, 2–6 Boundary Row, London SE1 8HN, UK

Chapman & Hall, 29 West 35th Street, New York NY10001, USA

Chapman & Hall Japan, Thomson Publishing Japan, Hirakawacho Nemoto Building, 7F, 1-7-11 Hirakawa-cho, Chiyoda-ku, Tokyo 102, Japan

Chapman & Hall Australia, Thomas Nelson Australia, 102 Dodds Street, South Melbourne, Victoria 3205, Australia

Chapman & Hall India, R. Seshadri, 32 Second Main Road, CIT East, Madras 600 035, India

First edition 1992

Typeset in 10/12 pt Times by Columns

Printed and bound in Hong Kong

ISBN 0 412 43910 7

A catalogue record for this book is available from the British Library

Library of Congress Cataloging-in-Publication Data

Mange, Maria A.
Heavy minerals in colour / Maria A. Mange, Heinz F. W. Maurer.
p. cm.
Bibliography: p.
Includes index.
ISBN 0–412–43910–7 (alk. paper)
1. Heavy minerals—Identification—Handbooks, manuals, etc.
I. Maurer, Heinz F. W. II. Title
QE364.2.H4M36 1989 89–9199
549′.1—dc20 CIP

Contents

Preface

Although some handbooks on the microscopic identification of heavy mineral grains are available, a comprehensive manual illustrated in colour has not been published until now. Because the appearance of minerals in grain mounts differs considerably from those seen in a thin section, a different approach is necessary for the identification of detrital grains. Coloured photomicrographs, showing their colour shades, pleochroism and interference tints, provide an excellent means of assisting recognition. As a number of mineral grains have similar optical properties and morphology, it is equally important to describe them verbally in detail, pointing out characteristic features and differences.

This book is intended primarily as a manual that describes and illustrates the transparent heavy minerals most commonly found in sediments. It is hoped that such a manual will be useful as a work of reference for students and research workers alike.

In Part I the concept of heavy mineral analysis is introduced and the relative significance of factors affecting heavy mineral assemblages is discussed. There are brief references to the commonly used laboratory methods and auxiliary techniques. It concludes with some examples of the application of heavy mineral studies.

Part II contains the descriptions of 61 transparent heavy mineral species, including those which are commonly authigenic in sediments. Positive identification of authigenic minerals is important to avoid confusion and to help recognition of diagenetic events. In the mineral descriptions considerable emphasis is placed upon detrital morphology and diagnostic features. Optical properties and characteristics are detailed, together with information on host rocks. Each mineral description is accompanied by one or more representative colour plates.

Acknowledgments

This book was completed while the authors were engaged in research at the University of Berne, Switzerland. I (MM) wish to express my gratitude to Professor A. Matter. His help and encouragement was instrumental in securing the book's publication. I am also indebted to N. H. Platt and S. D. Burley for their constructive criticism and valuable suggestions. Thanks are due to R. Oberhänsli for his help in the microprobe analysis and for reading some parts of the mineral descriptions. The manuscript benefited from fruitful discussions with P. A. Allen, Th. Armbuster, A. J. Hurford, A. C. Morton, Professor E. Niggli, K. Ramseyer and with several colleagues from the Geological and the Mineralogical-Petrographical Institutes of the University of Berne. F. Zweili has kindly provided assistance with the scanning electron microscope. We are grateful to G. Evans of Imperial College, London for allowing access to the heavy mineral collection of the late Professor H. B. Milner and to J. Waltzebuck of BEB Erdgas and Erdoel GMBH, Hannover for permitting the use of several heavy mineral grains from the Rotliegend of W. Germany. Aegirine, arfvedsonite- and dumortierite-bearing samples were kindly lent by A. C. Morton of the British Geological Survey, Keyworth.

We acknowledge the continued support and interest shown by Roger Jones, now of UCL Press.

Financial support was generously provided by the University of Berne, the Swiss Lottery Foundation, Shell Expro (U.K.) and BP Exploration, London.

Maria A. Mange-Rajetzky
Heinz F. W. Maurer

PART I

HEAVY MINERALS IN THE STUDY OF SEDIMENTS: PRINCIPLES AND PRACTICE

1 Introduction

High-density accessory mineral constituents of siliciclastic sediments are called **heavy minerals**. In their parent rocks they are present either as essential rock-forming minerals (e.g. amphiboles, pyroxenes, micas) or are accessory components, such as zircon, apatite, tourmaline, etc., occurring in a wide variety of rock types. Heavy mineral grains are seldom encountered in appreciable quantities in sandstone thin sections; their total quantity rarely makes up more than one per cent of the rock. In order to study heavy minerals effectively they need to be concentrated, and this is normally done by means of rock disaggregation and mineral separation, using liquids with densities of 2.89 (bromoform) or 2.96 (tetrabromoethane). High-density mineral grains sink in these liquids, hence the name 'heavy' minerals.

It is now more than 100 years since the study of heavy minerals was established. The first part of this century witnessed the enthusiasm of early researchers as they looked for and identified heavy minerals. During this time methods were refined and there was a proliferation of publications (e.g. Thoulet 1881, Retgers 1895, Mackie 1896, 1923a, Artini 1898, Illing 1916, Milner 1929, Boswell 1933, Edelman 1933, Baak 1936, Doeglas 1940). The popularity of heavy minerals until the 1940s stands in contrast to the general scepticism towards their value thereafter. The decline in popularity was due in part to the recognition that hydraulic effects cause selective sorting according to size, shape and density, and also to the realization that other phenomena may affect heavy mineral assemblages. For example, the occurrence of several mineral species is often grain-size dependent, and in addition post-depositional dissolution may often profoundly modify an initial mineral suite.

Largely as a result of these drawbacks, but also in part because of the arrival of more attractive and fashionable new techniques, many geologists regarded heavy mineral studies as uninteresting or useless. However, the technique has by no means been neglected. Throughout the 1950s and 1960s a number of researchers successfully used the study of heavy minerals in keynote works on a wide range of geological problems (van Andel 1950, 1955, van Baren & Kiel 1950, Hutton 1950, Wieseneder 1952, Koen 1955, Feo-Codecido 1956, Sarkisyan 1958, Sindowsky 1958, van Andel & Poole 1960, Baker 1962, Hubert 1962, Füchtbauer 1964, 1967, Hubert & Neal 1967, Gazzi & Zuffa 1970). These studies have greatly contributed to a better understanding of the factors important in modifying heavy mineral suites, and have opened the way for more sophisticated interpretations. The expansion of heavy mineral studies into hitherto neglected fields, the application of advanced analytical techniques and of mathematical data treatment – have all been important advances which have led to a rejuvenation of interest. Consequently, heavy mineral analyses are now more frequently employed and they yield highly informative results. This can be simply confirmed from scanning some of the latest volumes of the *Bibliography and Index of Geology*, published by the American Geological Institute. There are many studies in which heavy minerals play a conclusive role.

Our intention is to provide further assistance and encouragement for the study of heavy minerals and, we hope, to help in achieving more accurate results.

2 Heavy minerals in the study of sediments: their application and limitations

Where observation is concerned
chance favours only the prepared mind
Louis Pasteur 1822–95

Heavy mineral fractions in sediments are often composed of diverse mineral species, in which each grain conveys its own history. It is the sedimentary petrologist's task to decipher the message encoded in the assemblages and apply them for the purpose of:

(a) determining provenance: reconstructing the nature and character of source areas;
(b) tracing sediment transport paths: these are especially useful for complementing palaeocurrent analyses;
(c) mapping sediment-dispersal patterns;
(d) delineating sedimentary petrological provinces;
(e) outlining, and in suitable cases correlating, various sand bodies;
(f) indicating the action of particular hydraulic regimes and concentrating processes;
(g) locating potential economic deposits;
(h) elucidating diagenetic processes.

Heavy minerals are particularly useful in studies of sedimentation related to tectonic uplift, as the evolution and unroofing episodes of orogenic belts are faithfully reflected in their foreland sediments. Analysis of heavy minerals in foreland basin sequences may thus prove valuable in constraining the structural histories of both the basin and the tectonic hinterlands. Heavy mineral studies are also widely used in pedology, as they provide clues to soil formation.

From the moment the minerals are released from their host rocks a series of processes come into effect, operating until the assemblages are extracted from a sediment for study. The significance of factors likely to affect the reliability of heavy mineral analyses and their interpretation have been discussed and evaluted in many publications (Mackie 1923a, van Andel 1959, Blatt 1967, Hubert 1971, Pettijohn *et al*. 1973, Morton 1985a). The most important parameters are:

(a) Physiographic setting and climate of the source area. These factors largely control the pre-selection of mineral grains during host-rock weathering and thus determine the original input of heavy mineral species into a sedimentary system.
(b) Abrasion and mechanical destruction during transportation. These are related to mechanical durability of the grains themselves.
(c) Hydraulic factor. This operates during transportation and is controlled by the conditions of the particular hydraulic regime. Its effects result in selective grain sorting according to size, form and density.
(d) Post-depositional, diagenetic, effects ('intrastratal dissolution').

The last two parameters are especially important. The hydraulic factor decides which mineral grains will be deposited under certain hydraulic conditions. Post-depositional dissolution introduces a selection process of a different kind, namely the elimination of the less resistant grains, and it influences the ultimate heavy mineral assemblages of a particular sediment. Abrasion during transport appears to cause negligible modifications of heavy mineral suites (Russel 1937, Shukri 1949).

Hydraulic effects

Rubey (1933) presented the first conclusive paper on factors causing large variations in the relative mineral abundances of deposits that have been derived from the same source lithology. Rubey formulated the theory of **hydraulic equivalence** which states that grains of different sizes and densities, but of the same settling velocity, will be deposited under the same conditions. Consequently, the 'hydraulic equivalent size' is defined as a difference in size between a given heavy mineral species and the size of a quartz sphere with the same settling velocity in water. To resolve the problem of selective sorting, Rittenhouse (1943), in a study of sands from the Rio Grande, introduced the **hydraulic ratio** which he defined as '100 times the weight of a mineral in a known range of sizes, divided by the weight of light minerals of equivalent hydraulic size' (Rittenhouse 1943, p. 1743).

Subsequent investigations, however, have shown that the hypothesis of hydraulic equivalence is not universally applicable, and that in most sediments the heavy and light minerals are not in hydraulic equilibrium. Rittenhouse (1943), van Andel (1950) and Briggs (1965) attributed this discrepancy to limitations on grain-size distributions inherited from the host rocks, and particularly to the deficiency of larger size 'heavies', relative to larger 'lights', at source. van Andel (1950) noted that the relative availability of heavies varies consistently with transport distance; Lowright *et al.* (1972) later demonstrated that this was caused by hydraulic effects. In a study of the environment of Lake Erie these authors detected a systematic divergence from hydraulic equilibrium in that the heavy: light settling-velocity ratios displayed a systematic divergence from those expected from conditions of hydraulic equilibrium. They observed an apparent source influence in the river- and river-mouth sediments, but in longshore transport the effect of selective sorting appeared to be predominant. They concluded that deviation from the hydraulic equivalence was due not to source restrictions but rather to 'differential mineral transport', which resulted in the progressive elimination of heavies from the transporting system with increasing distance of travel.

Hand (1967) recorded that, once deposited, heavy minerals are more difficult to entrain than quartz. This theory was further developed by Trask & Hand (1985) in a study of the effect of hydraulic sorting of heavy minerals in longshore transport on the eastern shore of Lake Ontario. Trask & Hand introduced the concept of 'difficulty of entrainment' which implies that 'the smaller size of heavy minerals makes them less entrainable and less transportable than fall-equivalent lights'. This is partly explained by the tendency of the small heavy grains to 'hide' in the interstices of the larger light grains. Slingerland (1977, 1984) pinpointed factors such as 'selective grain sorting' during deposition and 're-entrainment processes', which ultimately exert most control on the final grain size of a deposit.

The dependence of marine concentrating processes on the energy of a particular environment was shown by Swift *et al.* (1971), Stapor (1973), Drummond & Stow (1979) and Riech *et al.* (1982). In the area of offshore Alabama and Mississippi, Drummond & Stow (1979) observed that the smaller and more dense heavy mineral grains (ilmenite, zircon, leucoxene, monazite and rutile) are concentrated in low-energy environments, whereas the larger and less dense heavy minerals (kyanite, tourmaline, staurolite and sillimanite) are found in areas of relatively high energy.

The influence of bed configuration on heavy mineral accumulations was noted by several authors. Hubert & Neal (1967) revealed that bottom topography largely controls sand dispersal patterns of deep-sea sands in the western North Atlantic petrologic province. Steidtman (1982) stressed that the 'hydraulic equivalence of grains is not only determined by conditions of deposition' but also by the 'specific style of grain motion and bed configuration'. The degree of hydraulic sorting of heavy minerals differs markedly between the upper and lower flow regimes. Steidtman advocated that 'sampling and analytical techniques should take such factors into account if a successful application of hydraulic equivalence is to be achieved, either in interpreting depositional processes or in reconstructing provenance'.

Results of recent studies, especially those which investigated the mechanisms of fluvial and alluvial placer evolution, have facilitated a better appreciation of processes operating in the entrainment, concentration and hydraulic equivalence relationships of light and heavy minerals (Slingerland 1977, 1984, Sallenger 1979, Komar & Wang 1984, Komar & Clemens 1986, Peterson *et al.* 1986, Komar 1987).

The effect of *grain shape* on the settling velocities of heavy mineral grains was investigated by Briggs *et al.* (1962). This work showed that the 'variations in shape' factor has an effect on the drag coefficient (a dynamic reaction between the fluid and the particle) of a similar magnitude to that caused by variation in specific gravity. Therefore 'sorting of heavy minerals by shape is as important as sorting by specific gravity'. The importance of hydraulic fractionation by shape in addition to density was also emphasized by Flores & Shideler (1978), expressed as 'shape-fractionation index'. This is calculated by the ratios of bladed and elongate minerals to equant minerals (e.g. pyroxene, hornblende, etc., versus garnet and epidote). Flores & Shideler showed that, during reworking of sediments on the South Texas Outer Continental Shelf by the Holocene transgression, the more equant epidote and also the heavier grains had a higher selection rate for permanent deposition.

Several attempts were made to evaluate quantitatively the effects of the hydraulic behaviour of heavy minerals on provenance determinations and correlation (for details see Hubert 1971 and Morton 1985a). However, the methods employed include time-consuming laboratory measurements as well as computation and they are not used in the general routine of heavy mineral analyses.

In studies of heavy minerals, the extent and possible effects of hydraulic sorting must always be considered. Its effect may be minimized by carefully planned sampling and selection of only those grain sizes typical for, and representative of, the sediment under study and the energy conditions operating during deposition.

Heavy minerals and grain size

A number of heavy mineral species have an affinity to certain grain sizes. This reflects their initial size in the source rocks, a factor controlled primarily by conditions of crystallization. The tendency of zircon to occur as small grains is often mentioned in this context. Other species, such as staurolite, kyanite, sillimanite, andalusite, topaz, and not uncommonly garnet, tourmaline and pyroxenes, often appear as fairly large fragments.

As both the size of the individual heavy mineral grains and the mean grain size of sediments are controlled by factors operating in the particular depositional environment, a uniform grain-size range, suitable for all heavy mineral analyses, cannot be established. Probably the only compromise is to assess the properties of the sediment in question and then to select, by means of careful examination and operational tests, a size range that will yield the most representative assemblage of heavy minerals and the maximum information. This practice is frequently adopted and two principal methods are used. The first limits the grades into one, often narrow, size fraction (e.g. Sindowsky 1938, 100–200 μm; Carver 1971, 125–250 μm; Gravenor & Gostin 1979, 75–150 μm; Morton 1985a 63–125 μm). This has the advantage of producing uniform observational conditions and eliminating apparent variations in heavy mineral proportions, caused by differences in grain size. In addition, using a narrow size grade helps to reduce the effect of hydraulic sorting. This approach is ideal for fine-grained, well-sorted sediments. However, in many cases it involves the risk of leaving undetected some diagnostic species, which are present only outside the selected size grade. This method may thus provide dubious results, especially in the case of coarse-grained deposits.

The second practice employs a wide size range or uses the entire heavy mineral 'crop' (e.g. van Andel 1950, van Andel & Poole 1960, 62–500 μm; Füchtbauer 1964, 63–400 μm; Galehouse 1967 and Rice *et al.* 1976, 62–2000 μm; and Milner 1962 and Pettijohn *et al.* 1973, the entire heavy mineral assemblage). These sizes will include all potentially useful species and, in addition, will provide a good characterization of the entire heavy mineral spectrum of a sample. The drawback of this approach is that the presence of fine and coarse heavy mineral grains within a single concentrate makes mounting and analysis difficult. When abundant coarse and fine grains are available, the problem can be tackled by splitting the heavy residue into two (or more) fractions. This may prove particularly beneficial for the analysis of poorly sorted sediments or for those which include detritus from source terrains, comprising complex lithologies and of various size grades. A far less frequently used but probably more effective approach is to convert relative percentages of heavy minerals to weight percentages (see p. 18).

Finally, it must be borne in mind, that for correlation, on either a local or a regional scale, only similar grain sizes and similarly treated and analysed heavy mineral suites should be used.

The chemical stability of heavy minerals

Dissolution processes of common rock-forming minerals have been extensively studied both in soils and in simulated laboratory experiments (among others, Raeside 1959, Wilson 1975, Berner 1978, Berner & Schott 1982, Nahon & Colin 1982, Anand & Gilkes 1984, Velbel 1984 (in soils); Huang & Keller 1970, 1972, Schott *et al.* 1981 (in the laboratory); see also Colman & Dethier 1986, Huang & Schnitzer 1986 and Tan 1986). These studies have investigated the rate and kinetics of mineral dissolution, and their results provide some insight into reactions between minerals and solvents, into dissolution processes through progressive etching, or into the conversion of one mineral into another mineral by replacement. Although these investigations were limited to surface or near-surface conditions, the processes probably take place in a similar style during burial.

The importance of post-depositional dissolution cannot be overemphasized. This can obscure provenance by eliminating informative unstable minerals, leaving a residue of ultrastable zircon, tourmaline, rutile and, often, apatite.

Details on the complex subject of *mineral stability* versus *dissolution* are beyond the scope of the present work. This is widely covered in the relevant literature, and the reader is referred to some key studies and reviews, such as Edelman & Doeglas (1934), Goldich (1938), Bramlette (1941), Pettijohn (1941), Dryden & Dryden (1946), Blatt & Sutherland (1969), Hubert (1971), Grimm (1973), Nickel (1973), Morton (1984b, 1985a). Several earlier studies on the stability of heavy minerals have been compiled and reviewed by Luepke (1984). Morton (1985a) summarized previous works on the chemical stability of heavy minerals and, basing his studies mainly on sequences from the North Sea, he established a realistic stability response of heavy minerals to acidic and alkaline geochemical environments (Table 2.1).

Thoulet (1913) was one of the first to note the decreasing complexity of heavy mineral suites with increasing geological age. Pettijohn (1941) considered that the general absence of many heavy mineral species in older sediments could be attributed to removal of unstable minerals by intrastratal solution. Pettijohn regarded age as a major controlling factor on the survival of minerals in sediments and this was expressed in the 'order of persistence' of heavy minerals (Table 2.1). Later studies (e.g. Wieseneder & Maurer 1958, Walker 1967, Walker *et al.* 1967, Grimm 1973, Friis 1974, 1976, Füchtbauer 1974, Scavnicar 1979, Friis *et al.* 1980, Morton 1982a, 1984b) dealing with subsurface formations which experienced acid or alkaline diagenesis have shed light on the conditions and processes that account for the gradual dissolution of unstable species. However, the overall behaviour of heavy minerals may be dependent on the chemistry of the individual mineral species in question. Mineral behaviour may also reflect different geological settings and different geochemical environments.

In general terms, the stability of a particular heavy mineral species can be determined from the pH of the geochemical environment, although caution is needed when using this simplistic approach since both Eh and ionic composition of pore fluids are also likely to influence heavy mineral stability. Unfortunately, there is still little data available on this subject at present. Heavy mineral assemblages appear to respond differently to extremes of acid environments (such as those typical in lateritic or humid-tropical weathering conditions) and alkaline environments (as typical in desert soils or as occurring in the saline brines associated with hydrocarbon reservoirs).

This difference of behaviour is well illustrated by the response of apatite and garnet under these extremes. During acidic weathering apatite is highly unstable, with the result that soils and rocks subjected to acidic leaching are usually depleted in apatite. The apatite dissolution effect is particularly strong in carbonate-free acid soils (Piller 1951). Lemcke *et al.* (1953) noted that in the presence of Ca^{2+} ions the solubility of apatite is reduced. However, it is commonly present in ancient sediments and in deeply buried rocks provided they were not subjected to acidic leaching. Morton (1984a, 1986) observed no dissolution of apatite in North Sea sequences even at burial depths exceeding 3800 m. In deeply buried Rotliegend sandstones apatite is present at even greater depth, where it frequently develops overgrowths (MM, unpublished data). As apatite is widespread in detrital sediments, absence of apatite from a sediment may be diagnostic of acid leaching either in the source area or at some stage during its history (Morton 1986).

Garnets are sensitive to acid leaching in the weathering profile (Bramlette 1929, Dryden & Dryden 1946), and their unstable nature in acids was confirmed in a laboratory experiment by Nickel (1973). The stability of members of the garnet group varies according to their chemistry. Humbert & Marshall (1943) found that melanite garnets, as well as iron and manganese garnets in the weathering products of basic rocks, were more weathered than the garnets from weathered granitic lithologies. Dana (1895) and Allen (1948) also noted that garnets with high ferrous iron content were particularly prone to disintegration. Morton (1987) observed that garnets with the highest calcium content were least stable and were dissolved at comparatively shallow depth in the Upper Palaeocene Forties Formation in the Central North Sea.

Recently, Velbel (1984) investigated the weathering process of almandine garnet in saprolite soils and in stream sediments in North Carolina. He demonstrated the different behaviour and dissolution of garnet in the studied environments as a response to controlling factors such as oxidizing potential and influences of organic and inorganic processes: garnet from the inorganic environment of oxidized vadose saprolite had a thick layer of gibbsite–goethite as weathering products, thus providing a layer which protected the grains from corrosion and their surface remained unetched. In soils where biological and biochemical processes prevail, and also in streams, a protective layer cannot form and a selective surface attack is shown by numerous etch pits.

The facetted forms of garnet have attracted considerable interest which provoked debates on their authigenic – (Simpson 1976) versus dissolution – (Gravenor & Leavitt 1981, Maurer 1982, Morton 1984b, Borg 1986) related origin. Hansley (1987) carried out an etching experiment on garnets, using organic acids, in order to reproduce the deeply facetted forms (common in the Morrison Formation in New Mexico) and to gain a deeper understanding of the process that produces facetted forms. Hansley provided petrological and experimental evidence, thus confirming studies that attributed the facetting to advanced etching, and discussed both the kinetics of etching and the factors controlling the dissolution processes in the Morrison Formation.

The effects of mineral dissolution can be observed in optical mounts, but the textural criteria for its recognition are most readily discerned using the scanning electron microscope (Hemingway & Tamar-Agha 1975, Setlow 1978, Berner & Schott 1982, Maurer 1982, Morton 1984b). Incipient dissolution produces small

Table 2.1 Order of persistence and generalized order of chemical stability of common heavy minerals. Minerals are arranged in order of increasing stability from top.

ORDER OF PERSISTENCE	GENERALIZED ORDER OF CHEMICAL STABILITY		
Modified after Pettijohn (1941)	Pettijohn et al. (1973),	Acid leaching (Morton 1985a)	Deep burial, saline or alkaline pore fluids, (Morton 1985a)
olivine sillimanite pyroxene sphene andalusite Ca-amphiboles glaucophane-riebeckite series epidote kyanite staurolite monazite xenotime apatite garnet zircon tourmaline rutile	very unstable olivine unstable hornblende actinolite augite diopside hypersthene andalusite moderately stable epidote kyanite garnet (iron-rich) sillimanite sphene zoisite stable apatite garnet (iron-poor) staurolite monazite ultrastable rutile zircon tourmaline anatase	olivine, pyroxene amphibole sphene apatite epidote, garnet chloritoid, spinel staurolite kyanite andalusite, sillimanite, tourmaline, rutile, zircon	olivine, pyroxene andalusite, sillimanite amphibole epidote sphene kynanite staurolite garnet* apatite*, chloritoid, spinel rutile, tourmaline, zircon * According to recent observations (MM) garnet and apatite appear to be more stable than chloritoid and spinel. The order of increasing stability from staurolite is therefore: staurolite chloritoid, spinel garnet apatite rutile, tourmaline, zircon

etch pits and mamillae on the surface of the affected minerals. As the process continues, the unstable grains, in accordance with their crystallography and chemistry, show facets, ragged edges, deep parallel grooves and furrows or become 'skeletal'. Grimm (1973, pp. 117 & 120) illustrated the progressive stages of dissolution for the most common heavy minerals. Of the numerous etch patterns, probably the 'hacksaw' or 'cockscomb' terminations of pyroxenes and staurolite, the 'frayed' edges of amphiboles, and the 'etch facets' of garnets are the easiest to recognize under the light microscope. However, caution is necessary before ascribing all these features to dissolution effects. Edelman & Doeglas (1932, 1934) noted that similar textures are often initially caused by mechanical processes and then enhanced by subsequent corrosion. Hubert (1971) argued that, in some cases, ragged outlines might reflect the original habit of mineral grains in the host rocks. Some irregular shapes are also produced by breakage during transport. In the course of our work we have seen typical 'hacksaw' terminations on freshly eroded pyroxenes in beach and river sands from southern Turkey and St Lucia. It therefore seems probable that grains may already possess inherited ragged outlines, superficially resembling etch features, before the time of deposition.

In deeply buried porous sandstone bodies, the dissolution process is most likely to modify the original mineral suite. Hubert (1962) suggested the **ZTR index** as an aid to the quantitative definition of mineralogical maturity in heavy mineral suites. The ZTR index is 'the combined percentage of zircon, tourmaline, and rutile among the transparent heavy minerals omitting micas and authigenic species'. This index is useful as a scale of the 'degree of modification, or maturity, of entire heavy mineral assemblages of sandstones'. In most greywackes and arkoses the ZTR index is about 2–39%. However, it usually exceeds 90% in orthoquartzites. This reflects the general increase in ZTR index with increasing geological age of the sediments as a result of the progressive dissolution of unstable minerals.

Early cementation may seal pore throats and isolate the rocks from circulating pore fluids, thus protecting the unstable minerals from dissolution. Concretions, thinly bedded sand bodies with frequent muddy horizons, and silty mudstones, usually have restricted fluid migration and thus contain better preserved heavy mineral suites (Bramlette 1941, Weyl & Werner 1951, Ludwig 1968, Blatt & Sutherland 1969, Morton 1984b, 1985a). Recently Blatt (1985) drew attention to mudrocks as potential and reliable rock types for provenance studies. Mudrocks are abundant in most sedimentary environments (they form 65% of the stratigraphic column) and the silt fraction of silty mudstones yields abundant heavy minerals and informative assemblages.

Oil infiltration into a sandstone halts or inhibits further diagenesis, and hence prevents or slows later mineral dissolution. Yurkova (1970) demonstrated that oil-saturated zones in the productive horizons of Sakhalin contain significantly more unstable accessory heavy minerals than the adjacent water-bearing formations. Morton (1982a) found similarly well-preserved unstable minerals in the oil-saturated Tertiary sandstones of the North Sea Balder Field and also attributed their preservation to oil inhibition of diagenesis.

Although intrastratal dissolution can profoundly modify initially diverse heavy mineral assemblages, an impoverished heavy mineral suite cannot be universally regarded as a consequence of diagnetic loss of the less stable species. van Andel (1959) argued that the absence of unstable heavies is in many cases due to parent rock compositions or to loss before deposition in areas of intensive weathering. van Andel stressed the prime importance of the geological setting: platform facies sediments, receiving detritus from cratonic sources, are generally characterized by stable assemblages; in contrast, sediments shed by orogenic sources contain diverse heavy mineral suites. In tectonically active settings, high relief, strong erosion and rapid burial provide conditions favourable for preserving an 'initial' composition of sediments faithfully reflecting the lithology of source lands.

Only a few studies of this kind have been carried out on pre-Cretaceous post-orogenic sediments, but their results appear to support van Andel's argument: the Permo-Triassic molasse of the pre-Ural region contains well preserved pyroxenes and amphiboles (Sarkisyan 1958). Within the clastic Upper Devonian series in the axial zone of the Pyrenees, Stattegger (1976) distinguished a 'western' heavy mineral province, typified by hornblende–zoisite–clinozoisite–epidote, which stood in sharp contrast with the adjacent heavy mineral provinces containing stable species only. The heavy mineral composition of these provinces was controlled by source-area lithology and not by post-depositional effects. Pyroxenes and hornblende are common constituents of the Devonian Old Red Sandstone (regarded as a post-orogenic molasse) of Scotland (Anderton *et al.* 1979, pp. 111–25). Cawood (1983) studied the compositions of detrital pyroxenes in early Palaeozoic rocks from the New England Fold Belt of Australia, and Weissbrod & Nachmias (1986) have reported the presence of unstable minerals from the late Precambrian Zenifim Formation of the 'Nubian Sandstone' in Israel, and from similar assemblages in southern Jordan.

To summarize, it is necessary to assess the local conditions prevalent in each individual study area before deciding whether (a) provenance controls or (b) dissolution processes have been ultimately responsible for the composition of the present heavy mineral suite. The latter is most readily recognized in unstable assemblages where, after partial dissolution, the presence of typical etch surface patterns and relics provide clear evidence of post-depositional modification of the assemblage. With increasing age or with greater depths of burial, intrastratal dissolution processes progressively diminish and ultimately eliminate the less stable species in accordance with their lower chemical stabilities.

Many heavy mineral suites contain only ultrastable minerals. The absence of unstables in such cases may be ascribed to: (a) source-area lithology (low-grade metamorphic terrains, carbonate rocks, mature polycyclic sediments – none of which contains significant proportions of unstable minerals); (b) pre-depositional or pre-burial loss (intensive chemical weathering in the catchment regions, low relief, slow deposition rate, advanced pedogenesis, corrosive groundwaters; any of these factors could lead to the breakdown of unstable minerals before sedimentation took place); (c) post-burial dissolution which would result in diagenetic loss of unstable minerals.

The relative importance of each of these processes may be evaluated by examining the light fraction of a sediment. Textural features (such as good sorting and extensive rounding) or particular compositional characteristics (such as absence of lithic fragments and paucity of feldspars) indicate reworking of already mature sediments, which as parent rocks are unlikely to furnish diverse heavy mineral assemblages. In contrast, the heavy mineral fractions of immature lithic arenites, arkosic sandstones, volcaniclastics, greywackes, etc., are generally rich in unstable heavy minerals. The absence or scarcity of unstable heavy minerals in such rocks may be explained in terms of effective intrastratal dissolution.

3 Methods

Sampling for heavy mineral analyses

The value of heavy mineral analyses largely depends on the accuracy of sampling, which consequently should be carefully and meticulously planned. To a large extent the purpose of the study and the nature of the depositional environment or environments in question will dictate the mode of sampling, but it is very important to design a dense sampling procedure and to collect unweathered material. Taking samples haphazardly will almost certainly yield meaningless and disappointing results. Samples should be taken at regular intervals as well as after changes of stratigraphy, lithology, facies, texture, flow regime, bed configuration, etc.

Though grain size may influence heavy mineral compositions, usually the fine-to-medium grained sands or sandstones yield the optimum heavy mineral assemblages. Siltstones and silty mudstones can often provide good results.

Almost all sands and sandstones contain at least some heavy minerals, even if 'dark spots', or streaks, indications of heavy minerals, are not apparent when investigating clastic sediments with the naked eye or under a hand lens. Heavy mineral rich layers are composed of hydraulically selected and -concentrated grains of normally high densities (in most cases opaques). They are important in recording specific hydraulic conditions (McQuivey & Keefer 1969, Cheel 1984), but are not representative of the overall heavy mineral content of a formation. Therefore, sampling these accumulations alone should be avoided.

Preparation of samples

The mass of a dry sample for heavy mineral analysis may vary between 100 g and 1000 g. From sands, immature sandstones or volcaniclastics, a sample of 100 g to 200 g will usually yield a sufficient quantity of heavy concentrate, but mature sediments and those which contain high proportions of cement necessitate a larger bulk-sediment sample size.

The following sequence of treatments usually precedes heavy mineral separation:

(1) disaggregation of coherent sediments, to liberate individual grains;
(2) acid digestion, to eliminate carbonates, and alternatively washing with water to dissolve soluble salts;
(3) removal of organic substances;
(4) freeing the grains from adhering clays or iron oxide coating;
(5) cleaning of bituminous or crude oil saturated samples;
(6) sieving to extract required grain sizes.

Size reduction, disaggregation and cleaning

The size reduction of many hard rock samples is best achieved by means of a laboratory roller-crusher or a jaw-crusher with aperture set to produce 5 mm chips. These in turn can be treated by chemical means until complete disaggregation occurs. For some samples a ceramic or iron percussion mortar and pestle are sufficient. The pestle should be applied with a crushing action (the force exerted vertically) to avoid destroying the original shape and size of the grains. Fine particles should be sieved out frequently to avoid over-grinding.

Mechanical disaggregation may introduce the danger of fracturing mineral grains. However, Henningsen (1967) experimentally demonstrated that the percussion method causes only negligible breakage of heavy mineral grains.

Acid digest is applied to dissolve unwanted carbonate material. The technique can facilitate the removal of calcareous particles from unconsolidated sediments and the release of other mineral grains from carbonate-cemented samples. Either dilute HCl (10%) or 10% acetic acid (CH_3COOH) is used. The former rapidly dissolves calcite but removes dolomite only with boiling. The serious handicap of HCl treatment is that it also eliminates phosphates. In view of the importance of apatite in fission-track dating and in studies of the thermal history of a formation, the HCl treatment should be avoided. Acetic acid leaves the phosphates intact, but a prolonged immersion is necessary to dissolve calcite (1–3 days, and subsequent boiling may still be required). Silica-cemented rocks may be disaggregated by *alkaline digest*, using KOH or NaOH solutions. However this treatment can seriously attack silicates.

Several other methods which serve to disaggregate coherent fine-grained rocks were described by Krumbein & Pettijohn (1938), Carver (1971) and Allman & Lawrence (1972), and many of these can be successfuly applied to sandstones. The *ultrasonic method* functions by breaking down the rock by high-frequency mechanical agitation. The technique is carried out using 5% sodium carbonate solution and a small quantity of wetting agent (e.g. Teepol) in a beaker which is placed in a sonic tank for 5–30 minutes. All these treatments should be followed by a thorough wash with water.

Grains may need to be cleansed of adhering clays or cement, and aggregates and flocculated particles should be dispersed. This is achieved by agitation using either a mechanical stirrer or ultrasound. The addition of a dispersant, such as sodium metaphosphate or sodium silicate, promotes efficient cleaning. Subsequently the fine suspension is decanted, the residue sands are poured onto a sieve of 0.063 mm, alternatively 0.053 mm aperture (US Standard mesh 230 or 270, British Standard sieves 240 or 300) and the fine particles are washed through by a spray of water.

For removing a pervasive iron oxide grain-coating Leith (1950) suggested oxalic acid and aluminium. A piece of aluminium is placed in a beaker containing a solution of 15 g oxalic acid and 300 ml water, the whole sample is added, and the mixture is boiled gently for 10–20 minutes. It may be necessary to add more oxalic acid to remove all iron. This treatment should be used cautiously as it attacks apatites, causing intense surface corrosion, alteration to a greyish-brown colour and a tendency towards aggregate polarization.

Samples containing fluid hydrocarbons and asphalt, or other bituminous substances, can be cleaned with petroleum distillates or with benzol, chloroform, ether, trichloroethylene or carbon disulphide. 'Soxhlett' equipment provides an effective method, in which crude oil is extracted from the samples with hexane (C_6H_{14}).

Certain formations contain abundant authigenic minerals which may settle in the heavy residue and thus mask the detrital population. Drilling mud also introduces contaminants into cuttings samples. Milner (1962, pp. 123–4) described the following methods for removing the most frequently occurring authigenic contaminants, but these techniques also destroy many detrital species.

Mineral	Treatment
anhydrite	hot, strong hydrochloric acid
baryte	concentrated sulphuric acid
gypsum	hydrochloric acid, ammonium sulphate or bromoform–benzol solution
pyrite	15% nitric acid, or warm hydrogen peroxide

Warm water dissolves gypsum and anhydrite. The heavy residue is placed in a fairly large beaker (500 ml) which is filled up with de-ionized water kept at about 30°C. The water needs to be changed once or twice daily and should not be allowed to evaporate, as this will promote precipitation. Anhydrite and gypsum usually dissolve within one week.

Sieving

Komar & Cui (1984) as well as Wang & Komar (1985) pointed out that sieving involves not only grade selection but also grain-shape selection. This is particularly relevant in the case of heavy minerals, which exhibit much more variation in shape than do light grains.

Before the main grading commences, unwanted finer particles are removed from the loose sands and disaggregated materials by a spray of water ('wet sieving') using a sieve of 0.063 mm or 0.053 mm aperture (230 or 270 US Standard mesh, 240 or 300 British Standard sieves). The selected grain-size range is extracted from the dry samples by standard sieving (Ingram 1971). If possible, the sieves should be agitated for 10 minutes using a mechanical shaker. A thorough cleaning of the screens after each sieving is imperative.

A test run may be necessary to reveal the grain sizes yielding the optimum results for a particular study. This can be done using a few representative samples, which are then sieved into narrow size grades and carefully weighed. Following heavy mineral separation and weighing the heavy residues of each size grade, a microscopic examination will reveal those fractions containing the optimum of heavy mineral assemblages. If appropriate, two or three narrow grades can be combined for a subsequent routine analysis.

Heavy mineral separation

General considerations

Concentration of heavy minerals is performed by means of high-density liquids. There is a considerable difference in densities between the framework constitutents and heavy accessory minerals of a sediment. Upon immersion of a sample in a liquid of an intermediate density 'sink' (higher density) and 'float' (lower density) fractions are produced. These are commonly called heavy and light fractions respectively.

The acquisition of reliable heavy mineral data largely depends on minimizing laboratory errors. The choice of techniques used will also have an effect on the quality of the results obtained. Rittenhouse & Bertholf (1942) compared the effectiveness of gravity settling and centrifuge separation. They observed that the weight percentages of heavy mineral concentrates obtained by the two methods differ significantly, but the number frequencies of the individual heavy minerals are the same in both cases. Recently, Schnitzer (1983) presented startling results of his laboratory experiment on Triassic Buntsandstein samples. He found high fluctuations in heavy mineral percentages from samples separated simultaneously, using either different liquids (bromoform or tetrabromoethane) or different separation techniques (gravity or centrifuge methods). Variations also occurred when different settling times or differing densities of bromoform were used. The causes of these fluctuations are not completely understood. However, Schnitzer's work underlines the importance of considering the possible influence of separation technique on heavy mineral data.

High density (highly toxic) liquids normally used for separation are:

	Density at 20°C
bromoform (tribromoethane)	2.89
tetrabromoethane (acetylene tetrabromide)	2.96
methylene iodide (di-iodomethane)	3.32
Clerici's solution	4.24

The latter is a high-density aqueous solution of thallous formate–malonate. It is of particular use for the concentration of some of the 'heavier' minerals such as garnet or zircon.

Washing liquids suitable for removing the heavy liquids from the grains are: carbon tetrachloride, benzene, alcohol (ethyl alcohol or methylated spirit) and acetone. Because of their low toxicity, alcohol and acetone are most frequently used.

Regarding diluents, IJlst (1973) recommended isoamylacetate as a stabilizer and diluent for bromoform, and orthodichlorobenzene for methylene iodide. IJlst's tests indicated that use of a weight proportion of 1% isoamylacetate to the bromoform will produce a stable solution which has a constant density (2.82) and will not decompose during use or storage. He obtained similar results with methylene iodide and orthodichlorobenzene solutions.

Recently the non-toxic sodium polytungstate ($3Na_2WO_4.9WO_3.H_2O$) has been proposed for heavy liquid separation to replace the highly toxic heavy liquids (Callahan 1987). Sodium polytungstate is available in either liquid or powdered form from SOMETU (Falkenried 4, D–1000 Berlin 33, West Germany). It is water soluble and the mineral separates can be cleaned with distilled water, though with some difficulty because of the high viscosity of the liquid.

In geological laboratories two basic techniques are employed for 'standard' heavy liquid fractionation: (a) gravity settling or funnel separation, and (b) centrifuge separation. The latter is best for separation both of very fine (<63 μm) and of sand-size sediments. This technique was already advocated in the mid-1920s, but workers consistently encountered difficulty in achieving complete and uncontaminated recovery of the two separates from the centrifuge tubes. The introduction of 'partial freezing' (Fessenden 1959) largely alleviated this problem and vastly improved the centrifuge separation technique so that it is now in very widespread use. Fessenden (1959) employed solid carbon dioxide (dry ice) to the lower part of the centrifuge tube, permitting freezing of the separate containing the heavy fraction. Scull (1960) suggested liquid nitrogen as an alternative freezing agent.

Centrifuge methods which omit freezing the heavy liquid employ various modifications to the centrifuge tube design (e.g. constricted tubes or a 'tube within tube' arrangement). In such cases the tubes are equipped with a closure rod or stopper to ensure a clean recovery of the heavy fraction (Hutton, in Tickell 1965, Allman & Lawrence 1972 and Gautier & von Pechmann 1984).

Gravity separation

All separations must be carried out in an efficient fume cupboard. Rubber gloves are essential to prevent contact with the liquid.

Figure 3.1 illustrates the arrangement of equipment used for separation by gravity settling. The heavy liquid is filled in the separating funnel, the dry and weighed sample (maximum 10 g) is then added to the liquid which should be stirred to ensure that the grains are thoroughly wetted. Grains adhering to the stirring rod or the side of the funnel are removed using a jet of heavy liquid. Heavy minerals will accumulate in the bottom of the funnel above the pinch clip. When no more grains sink (normally after 6–8 hours separating time), the pinch clip is opened slowly, thus allowing the heavy fraction to pour onto the filter paper in the lower funnel. The pinch clip is closed, leaving a layer of clear liquid below the light fraction. A new funnel, with filter paper, is placed under the separating funnel. The light fraction is then drained into the new funnel. Subsequently the wall of the separating funnel is rinsed with a jet of heavy liquid in order to remove any adhering grains. Both fractions are washed thoroughly with alcohol or acetone and set aside to dry. The used heavy liquid and washings are collected in labelled bottles. The heavy liquid can be re-used. After drying, the heavy fraction is weighed and bottled in glass vials.

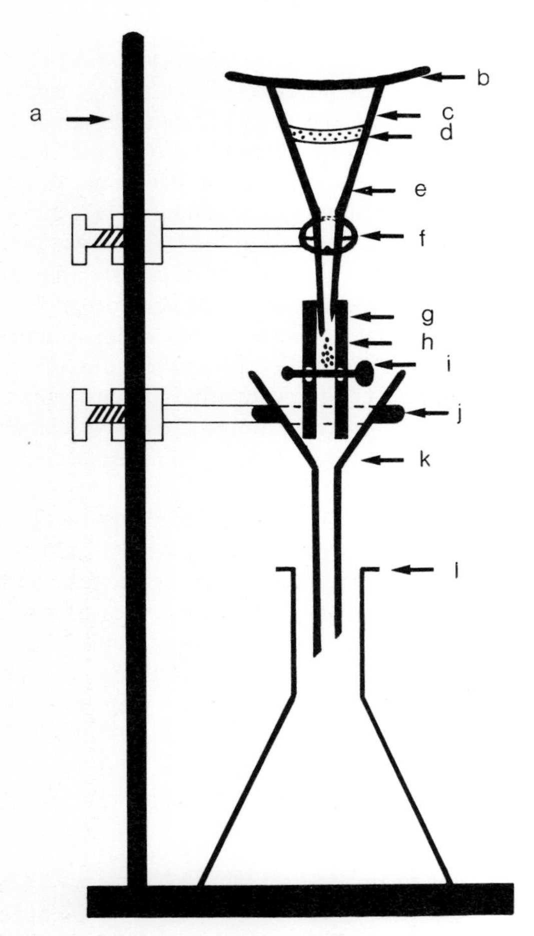

Figure 3.1 Arrangement of equipment for heavy mineral separation by gravity settling
(a) Retort stand, (b) watch glass, (c) separating funnel, (d) position of light fraction, (e) heavy liquid, (f) funnel support, (g) rubber tube, (h) position of heavy residue, (i) pinch clip, (j) filter funnel support, (k) filter funnel, (l) collecting bottle.

Centrifuge technique
Allow 10–15 minutes in the centrifuge for the separation of fine-grained sediments. Separating times can be shorter for coarser sediments. Centrifuge speed is usually set between 2000 and 3000 rpm.

The sample volume should be in proportion to the capacity of the centrifuge tube. The best results are achieved when the ratio of sample to heavy liquid is 1:10. Extreme size differences should be avoided. Separation is most accurate when the particles are of fairly uniform granulometry. The most commonly used centrifuge tubes are shown in Figure 3.2.

Partial freezing method Dry and weighed samples are poured into the centrifuge tubes, heavy liquid is added and the mixture is stirred or shaken thoroughly to wet all the grains. The stirring rod and the walls of the centrifuge tube are washed with heavy liquid, then the

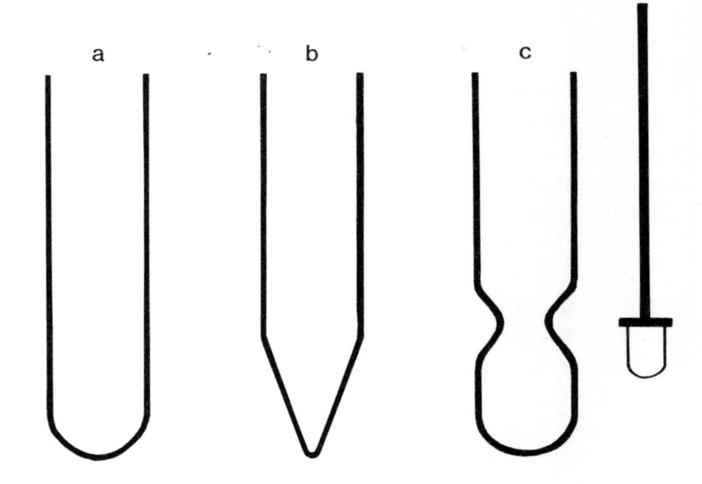

Figure 3.2 Commonly used centrifuge tubes
(a) Standard glass tube, (b) glass tube with tapered end, (c) constricted, ('Hutton') tube and closure rod.

centrifuge tubes are balanced. While the centrifuge is in operation, two sets of funnels and filter papers are arranged. When the centrifuge has stopped, the tubes are carefully lifted and placed into a test-tube rack over 'dry ice', or held by crucible tongs, lowered onto liquid nitrogen or liquid air. The lower part of the liquid, which contains the heavy fraction, will freeze within seconds. The liquid above this frozen plug is poured into a funnel and the walls of the centrifuge tube are rinsed carefully with a jet of heavy liquid from a squeeze bottle held in inverted position. The frozen plug is emptied into another funnel and the centrifuge tube is rinsed with heavy liquid to recover all heavy grains. The separation process is completed in a manner similar to that used for the funnel separation method.

'Tube within tube' method (after Gautier & von Pechmann 1984) This separation involves no freezing. It requires standard 100 ml centrifuge tubes fitted with a tapered inner tube (Figure 3.3). The inner tube has a collar attaching it firmly to the rim of the centrifuge tube. The tapered end of the inner tube has an opening of 2 mm in diameter which can be closed with a PVC, polyethylene or Teflon rod. The filling level of the heavy liquid is marked on the tubes. It is important to regulate the level so as to prevent lifting the inner tube by hydrostatic pressure.

The inner tube is filled with heavy liquid near to the level mark then the closure rod is inserted. The sample is poured into the inner tube and the mixture is stirred with a glass rod which is then rinsed with heavy liquid. The closure rod is extracted and rinsed from adhering particles. The centrifuge tubes are balanced and the centrifuge is started. The light fraction will float within

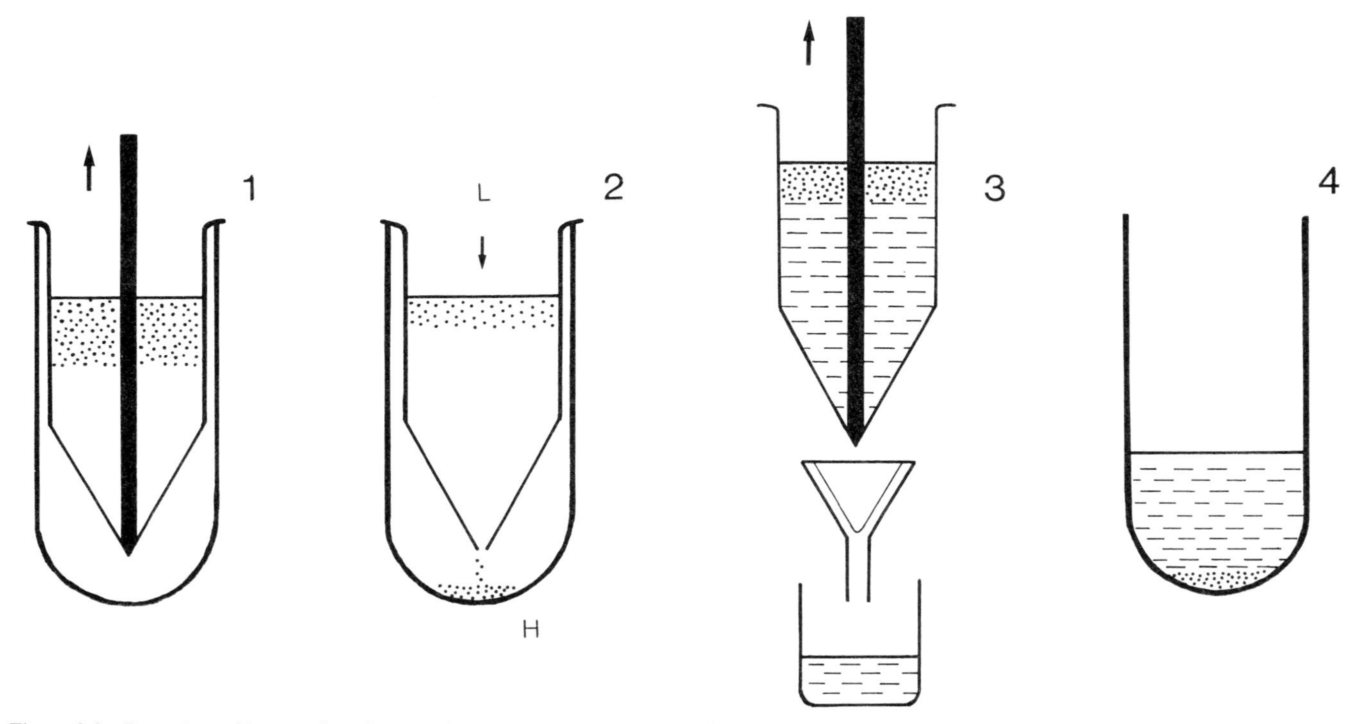

Figure 3.3 Procedure of heavy mineral separation by 'tube in tube' method (after Gautier & von Pechmann 1984)
(1) Arrangement of tubes, heavy liquid and bulk sediment before separation; (2) after centrifuge separation. L: light fraction, H: heavy fraction; (3, 4) recovery of the light and heavy fractions.

the inner tube, but the heavy grains will escape through its opening and settle on the bottom of the outer tube. After the centrifuge has stopped, the closure rod is carefully inserted. The inner tube is then removed and the light fraction is decanted into a funnel. The heavy fraction is emptied into another funnel and the centrifuge tube is rinsed of remaining grains with a jet of heavy liquid.

Miscellaneous separations

Panning may be used to reduce the initial volume of a sample. Hutton (1950) suggested panning in the field as a method to pre-concentrate loose sands and soil material. Panning is usually followed by heavy liquid separation to gain a clean concentrate.

The *concentrating* or *shaking table* is used in mineral processing and in cases where minor accessories such as minerals for radiometric dating or chemical analyses need to be segregated from large bulks of material. Recently, Stewart (1986) proposed its application for the routine separation of heavy minerals. Bearing in mind the health hazards of heavy liquids this technique may be viable as a safer alternative, but it may not be efficient for the separation of small volumes.

Magnetic separation of sedimentary minerals is often carried out on heavy mineral fractions for the purpose of segregating a certain mineral species or group of minerals based on their different magnetic susceptibilities. Magnetic separators are usually supplied with practical notes for application and tables showing the magnetic susceptibilities of common minerals (e.g. Hess 1956). Tables indicating magnetic susceptibilities and the best extraction ranges were compiled by Rosenblum (1958), Tickell (1965, with reference to Flinter 1959), Parfenoff *et al.* (1970) and Allman & Lawrence (1972). This technique is particularly useful for concentrating garnets, pyroxenes and amphiboles for chemical analysis, or apatite and zircon for geochronology.

Details of the above techniques together with the description of several additional methods developed for concentration or segregation of minerals with special properties, were described by Krumbein & Pettijohn (1938), Milner (1962), Tickell (1965), Parfenoff *et al.* (1970), Carver (1971) and Muller (1977).

Preparation for optical analysis

Splitting of heavy mineral fractions

If a heavy residue is too large to be mounted evenly on a microscope slide, splitting becomes necessary and this has to be performed precisely; many analysts do not realize the seriousness of errors involved in careless splitting. Sprinkling part of the residue from a vial onto a microscope slide induces grain sorting. Hutton (1950) advised strongly against 'the common practice of taking samples with the point of a pen-knife'.

Microsplitters (small devices used for splitting) are efficient means of producing small portions for grain mounts. Krumbein & Pettijohn (1938) described several types. When a microsplitter is not available a method suggested by Hutton (1950) serves as a good alternative. A small funnel with an upper aperture of about 25 mm diameter and a horizontally cut lower aperture is placed onto a large microscope slide, or onto a piece of sheet glass. The heavy fraction is poured into the funnel, which is tapped gently to shake down any adhering grains; the funnel is then lifted slowly off the plate. The cone of grains, formed this way, can be divided by quartering with a razor blade.

Grain mounts

For temporary mounts various immersion liquids are used, the most commonly used being clove oil, (n = 1.53), monobromnaphtalene (n = 1.658), cedar oil (n = 1.74) or methylene iodide (n = 1.74).

Permanent partial embedding and liquid immersion method The principle of this technique is to stick the grains onto the microscope slide and then to apply successive refractive index liquids to the same assemblage. After inspecting the minerals in a liquid of a particular refractive index, the slide can be washed. Another liquid with a different refractive index may then be applied. This method is helpful either for routine analysis or as a special aid for distinguishing between minerals of similar appearance but different RI.

Various agents may be employed to adhere grains to the microscope slide. Spencer (1960) recommended Collodion (n = 1.505). This is only successful with silt-size particles. Nayudu (1962) suggested gelatine for sticking the grains onto the glass slide, and Gazzi *et al.* (1973) used gum arabic. The latter is commonly employed by Italian sedimentary petrologists for routine heavy mineral analysis (G. G. Zuffa, personal communication 1986), and in the preparation of the grain mounts for the illustrations in this book, I (MM) also used gum arabic before applying the mounting resin.

A mixture of 1:4 gum arabic and distilled water is spread onto a microscope slide and allowed to dry for about three hours. Grains are sprinkled evenly onto the surface of the dry gum arabic film. To fix the grains the surface is moistened by breathing onto it and then dried for a few minutes under a lamp. Immersion liquids may then be applied which can be subsequently removed with acetone. A single slide may be used with a number of successive immersion liquids.

For *permanent mounts*, various mounting media are available and the type employed depends on the preference of the individual analyst. The classic resin, Canada balsam (n = 1.538), has been largely replaced by synthetic resins, but many workers still prefer to use it and it is easily obtainable. The advantage of liquid Canada balsam is that it has a molassic consistency which is excellent for 'crystal rolling', enabling examination of grains in different orientations and thus greatly assisting the identification of difficult minerals. If necessary, grains can be easily recovered later with a sharp needle for auxiliary analyses. Retrieved grains are placed in a drop of benzene or xylene for cleaning. Mounts made in liquid Canada balsam have to be stored flat on a tray. Alternatively, the balsam can be hardened by heating at 115°C.

Synthetic resins with low refractive indices are Eukitt (n = 1.54), Lakeside (n = 1.54), Permount (n = 1.567) and Entellan (n = 1.490–1.500). The latter is a new embedding agent intended to replace Caedax and produced by Merck (Darmstadt, West Germany).

Many workers favour high-index resins, as these permit distinction beween species by comparing the refractive indices of the minerals with higher and lower refractive indices than the resin. However, it is difficult to study lower-index minerals, especially those that are colourless, as their outline is almost invisible in a high-index resin, and surface textures and cleavages are hardly detectable.

A new range of mounting media are available in the 'Cargille Meltmount' series, produced by Cargille Laboratories Inc. (55 Commerce Road, Cedar Grove, New Jersey, 07009–1289 USA). These are specially formulated optical-quality thermoplastics and are ideal for permanent mounts. Meltmount 1.539 is recommended as a replacement for Canada balsam. Of the high-index series, Meltmount 1.662 is a direct replacement for Aroclor 5442 and Meltmount 1.704 for Naphrax. In addition there is a series of other Meltmount media with refractive indices: 1.582, 1.605 and 1.680. All these products are fluid at 65°C, require no curing time and are thermally reversible for particle retrieval or re-orientation. They are soluble in toulene, acetone, ethyl ether and methylene chloride. Mounts are made on a hot plate (this should be set first at about 80°C to melt the resin, then reduced to 65–70°C for the mounting procedure). Grains are placed on the microscope slide, covered with a cover glass, and a few drops of Meltmount are applied with an eyedropper until the grains are surrounded. Meltmount should not be overheated, as bubbles form easily.

Piperine (n = 1.68) was suggested for use in grain mounts by Martens (1932).

Mounts in epoxy casting resins Methods employing this technique were described by Chatterjee (1966), Middleton & Kraus (1980) and Leu & Druckman (1982). We use Araldite D and Hy 956 hardener in a 5 : 1 ratio and prepare the mounts on a hot plate to speed up curing time. The hot plate is heated to 90°C, a drop of Araldite is placed on a microscope slide and the grains are sprinkled onto it and then stirred in a circular motion with a dissecting needle to ensure even grain dispersal. The temperature is raised to 120°C, causing the hardener to evaporate in about two minutes. A mount prepared this way is very durable and is easy to polish. When only a few, coarser grains are available, they may be mounted on a microscope slide with double-sided adhesive tape. A mould is made by placing a plastic ring around the grains, filled with Araldite and then treated as above. (Note: some Araldite is birefringent and not suitable for ordinary grain mounts; therefore caution must be taken to use non-birefringent Araldite).

Heavy mineral grain mounts made using this technique are suitable for the identification of associated composite grains and rock fragments and, if necessary, for the staining of carbonates or sulphates. Uncovered and polished, these mounts can be used for electron microprobe or cathodoluminescence analyses.

When preparing grain mounts, it is essential to distribute the grains evenly on the slide, avoiding overlapping and dense packing of grains. Even dispersal and preferred grain orientation are promoted by gently sliding the cover glass backwards and forwards with plastic tweezers or a pencil eraser while the mounting medium is in a liquid state.

Microscopic identification

Many publications deal with the identification of minerals by their optical properties (Heinrich 1965, Tröger 1969, Phillips & Griffen 1981, Nesse 1986, Pichler & Schmitt-Riegraf 1987), and discussion of the topic is unnecessary here. However, as most optical mineralogy textbooks convey information and illustrations almost exclusively for the investigation of minerals in thin section, it is important to point out some major differences arising during examination of the same mineral species in the form of detrital grains.

In grain mounts, in contrast with thin sections, all grains are surrounded by the mounting medium, and relief is prominent. The technique of partial embedding (p. 16) is especially advantageous for a good approximation of the refractive indices. This is carried out either by using the familiar Becke method or by 'oblique illumination', otherwise known as Schroeder–van-der-Kolk test (see Muir 1977, pp. 45–7). Van Hilten (1981) proposed the technique of 'oblique observation' which needs no out-of-focus manipulation of the microscope. It is based on the phenomenon that light rays, departing from an 'interface' (i.e. the junction between two grains or between a grain and a mounting medium) are denser above that which has higher refractive indices. Looking obliquely into the microscope by moving the head slightly will cause blocking of the higher- (or lower-) intensity part of the light beam. In the former case the result is a darker interface as it receives less illumination, whereas a brighter contact will appear with the opposing movement. Observation is promoted by slightly closing the substage diaphragm.

When the precise measurement of the optical constants of individual anisotropic mineral grains by the immersion method is necessary, the spindle stage proves a valuable tool (see Bloss 1981, and Nesse 1986, p. 294).

Detrital minerals often appear in diverse forms and take up various orientations upon mounting. Those with a good cleavage or parting settle in the immersion medium on their best cleavage, but many species, known to show good cleavages in thin section, fail to display any in grain mounts. Grains are usually thicker than their counterparts in thin section, hence colour, pleochroism and interference tints are more intense and could be misleading. Preferred particle orientation, poor preservation or strong interference colours can inhibit the determination of the optical character and elongation. Irregular habit and lack of visible cleavages may hinder the measurement of the extinction angle. Though the above factors may cause confusion, most detrital minerals have fairly consistent diagnostic features which assist identification (see Part II).

Grain counting

To determine the relative abundance of heavy minerals in a sample, three different point counting methods are used (van Harten 1965, Galehouse 1969, 1971):

(a) *Fleet method* (Fleet 1926): All grains are counted on the microscope slide, and relative abundances are calculated to give number percentages.

(b) *Line counting*: The microscope slide is moved by means of a mechanical stage along linear traverses ('lines') and individual grains which are intersected by the crosshair are identified and counted. The results are number frequencies. This method is grain-size sensitive, as the 'lines' intersect the larger grains more frequently than the smaller ones, thus leading to a distortion in favour of the larger grains.

(c) *Ribbon counting*: This technique involves randomly selecting ribbons (bands) within the microscope slide. Grains within the bands are counted while the slide is manoeuvred by a mechanical stage. The width of the ribbons can be measured conveniently using a micrometer eyepiece. The results of ribbon counting are independent of grain size and they yield number frequencies. Judging from references to counting methods, ribbon counting seems to be the most popular.

When analysing heavy mineral assemblages the non-opaque, non-micaceous suites are normally counted excluding authigenic minerals. Some researchers count opaque grains and 'alterites' (weathered unidentifiable or composite grains; van Andel 1950) separately and express their abundance as a percentage of the total grains counted. In assemblages dominated by the volume of a prominent mineral (e.g. garnet or apatite), this species may be counted separately, similarly to the above (Füchtbauer 1964), thus allowing more accurate estimates of rare but diagnostic minerals. The number of grains counted is usually recalculated to equal 100 % and the abundance of individual minerals is expressed in percentages by number.

The precision of heavy mineral analysis is largely a function of the number of grains counted on a microscope slide. Dryden (1931) graphically illustrated the reliability of heavy mineral data in relation to the number of grains counted and observed a rapid increase in accuracy with increasing number of counts up to 300. Above this number the probability of inaccuracies is greatly decreased. In a test on Rhine sediments, van Andel (1950) counted separate batches in multiples of 100 between 100 and 600 grains. He found some variations in the proportions of the most abundant minerals, though these variations never exceeded 5 %. Rare species were encountered only when 200 or more grains were counted. In view of such experiments it is necessary to count at least 200 non-opaque grains to arrive at a reasonable estimate of mineral proportions and also to detect rare but often diagnostic species reliably.

The efficiency of the analyses increases considerably, when other mineral characteristics, such as morphology, colour, etc., are noted at the same time as the grains are counted. For example, distinction between euhedral and rounded forms of zircon, tourmaline and apatite may permit the recognition of different sedimentary histories (see below).

Some authors have attempted to convert number percentages to weight percentages (e.g. van Andel 1950, Ludwig 1955, Hunter 1967, Norman 1969, Mange-Rajetzky 1983; this method was also referred to by Füchtbauer 1974 and Brix 1981). Conversion calculations are based on the densities of the individual minerals and on the total weight of the heavy fraction. Weight percentages provide reliable values for the overall frequencies of a mineral species in a sample and are especially useful when several size grades are analysed. For accurate results all species should be counted, making this a very time-consuming approach.

Study of varietal types

Brammal (1928) pointed out that 'the *varietal* features of a species in a detrital assemblage may be of greater significance than the mere presence of that species'. Indeed, during the history of heavy mineral analyses many studies have proved that the value of results has increased considerably when the various types of a particular mineral (or of a mineral group) were distinguished and counted (i.e. the varietal types). Recording of varietal features may facilitate the successful correlation or distinction of sand bodies, and may also yield clues about progressive recycling of a particular formation into successively younger deposits. The recording of varietal characteristics will also permit more refined reconstruction of source-area lithology and provenance. Focusing on one particular mineral or mineral group helps to minimize the effects of the hydraulic factor and of other (e.g. diagenetic) factors acting to modify the members of an original heavy mineral suite. Varietal studies can be conducted on criteria of physical properties, such as mineral colour, inclusions, striations, twinning, overgrowth, zoning, etc. These characteristics may be studied under the polarizing microscope or alternatively under cathodoluminescence. Grain size and the degree of rounding may also be recorded.

The use of physical properties may be applied to all types of sediments, provided that the selected minerals are sufficiently abundant. It is necessary to count 75–100 grains of a particular species, recording the informative varietal types in separate categories. Zircon is an ubiquitous detrital mineral and its variations in morphology, colour and textural characteristics all lend themselves to successful varietal counts (see details under zircon in Part II). For instance, the proportions of zoned and unzoned zircon grains permitted the correlation of members of the Karroo System in southern Africa (Koen 1955). The colour and morphological criteria of zircons in the Morrison Formation of

New Mexico assisted in distinguishing between zircons sourced by Precambrian and Mesozoic intermediate intrusive rocks and those derived from Tertiary rhyolitic airfall tuffs (Hansley 1986).

Tourmaline has many varietal types. However, these are not always consistent, even within a particular host rock formation, and hence provenance reconstruction on this basis alone should be avoided. For example Staatz *et al.* (1955) found significant chemical and colour variations in tourmalines between the walls and core of a single pegmatite. In contrast, Power (1968) and Henry & Guidotti (1985) showed that the chemical characteristics of tourmaline compositions can be useful petrogenetic indicators under particular conditions. Krynine (1946) summarized the use of detrital tourmaline in tracing sediment provenance. Beveridge (1960) was able to distinguish three major parent-rock types in the Lower Tertiary Formations of the Santa Cruz Mountains o[illegible]rnia on the basis of tourmaline shapes and c[illegible]. In a study conducted by MM on sediments from Tunisia, slender green prismatic tourmalines with a darker nucleus were reliable markers of several sedimentary formations with a common parentage.

The crystal form, colour, characteristic strings of inclusions and pleochroic cores of apatite are all informative properties for varietal studies. Fleet & Smithson (1928) traced the source of several deposits in Wales and the British Midlands to the Leinster granite with the aid of cored and dark apatites. Varieties characterized by parallel strings of dark inclusions were useful index minerals in sediments derived from a common source in MM's study in northern Tunisia.

Rutile, which is found in a number of varieties in sediments, is also a suitable candidate for varietal studies (Galloway 1972, Riech *et al.* 1982).

Inclusion-free staurolites indicated a Catalanides source, whereas those which contained abundant carbonaceous or quartz inclusions indicated a Pyrenean parentage for the Oligocene sediments of the eastern Ebro Basin (Allen & Mange-Rajetzky 1982). Hansley (1986) also noted a change of staurolite varieties between the lower and uppermost part of the Westwater Canyon Member of the Morrison Formation, New Mexico.

A more advanced approach involves mineral chemistry. The varieties of a mineral species or group may be distinguished by their chemical composition. Morton (1985a) summarized the results of several studies by this method. Pyroxenes, amphiboles and epidote minerals are most suitable for varietal studies by chemical means, since their compositions show great variability through solid solution series; but, owing to the unstable nature of these minerals, their occurrence is usually limited to younger sediments. Morton (1985b) advocated study of the compositional varieties of garnets as they are relatively stable, faithful indicators of their mode of origin and are widespread in sediments.

Advanced auxiliary techniques

Though the petrographic microscope remains the fundamental tool in heavy mineral identification, in recent years additional advanced techniques (such as X-ray diffraction, electron microprobe, scanning electron microscopy and cathodoluminescence) have also been employed, further assisting optical identification. Use of these techniques provides information on chemical composition, structural and textural characteristics, and thus considerably increases the accuracy of heavy mineral analyses, permitting more confident interpretations to be made. The techniques are applicable to detrital minerals just as they are to their counterparts in 'hard rocks'. However, sample preparation may be laborious.

Sufficient quantities of a particular mineral species can be concentrated either by conventional separation techniques or by hand picking. As in most cases the number of available grains of a particular species is limited, hand picking is often the only approach possible. McCrone & Delly (1973) and Muller (1977) described several useful methods for manipulating minute particles. Fine needles are important for hand picking. They can be made from dissecting needles by grinding them to a finer point, or from 24–6 gauge tungsten wires and a sodium nitrite ($NaNO_2$) stick (McCrone & Delly 1973). 'Tungsten reacts exothermically with sodium nitrite to form sodium tungstate. This serves to chemically etch the tungsten and, when applied to a wire, a fine-tip needle is quickly produced.' A stick of sodium nitrite is held over a Bunsen burner and, when the needle is red hot and the tip of the stick begins to melt, the wire is gently struck through the sodium nitrite melt, thus producing the fine-tip needle.

It may be useful to include a few hints about isolating grains, as this seemingly difficult process might deter workers from applying further tests. Different mineral species can be recognized, and easily selected, under the stereobinocular microscope by means of their colour or morphology. Colourless, irregular and rare grains are easier to detect under the petrographic microscope when immersed in a drop of alcohol, distilled water, or in partial embedding and immersion liquid mounts. The location of the grains may be marked with a fine marker pen or by placing a small chip from a crushed cover glass over it, so that it may still be found after the slide has been transferred to the binocular microscope, from where the grains may be more easily retrieved. The selected grains are transferred onto a microscope slide and a drop of distilled water or alcohol is added. Evaporation of the fluid causes the grains to adhere to the slide and helps to prevent loss. Further preparation will depend on the nature of the analysis planned.

X-ray diffraction

X-ray diffraction (XRD) aids mineral identification by providing data on crystal structure. XRD can also be useful for identifying opaque grains (Harrison 1973) and for obtaining a general knowledge on bulk heavy mineral composition (Pryor & Hester 1969). The X-ray method requires finely ground (1–10 μm) crystal powder. Powder cameras are used because the quantities of material available are usually minute. For further details of the X-ray technique, the reader is referred to Guinier (1964), Klug & Alexander (1974) and Zussman (1977).

X-ray fluorescence spectrometry

X-ray fluorescence (XRF) spectrometry relies upon bombardment and excitation of elements by means of a primary X-ray beam; this results in the emission of X-rays with characteristic wavelengths for each element. This radiation is detected by an X-ray spectrometer and is then used to identify and estimate the concentration of elements in samples (Norrish & Chappell 1977).

XRF may be used on the whole heavy mineral fraction and may prove advantageous for distinguishing between similar lithological units and petrological provinces by the trace element contents of their detrital minerals (Riech *et al.* 1982). For more information on this technique, see Norrish & Chappell (1977).

Electron-probe microanalysis

The electron microprobe plays an essential part in identifying unknown or ambiguous grains, and in distinguishing between individual members of a mineral group or series by providing the necessary data on crystal chemistry. Several recent studies have proved the value of the electron microprobe in heavy mineral analyses. Knowledge of the mineral chemistry of detrital pyroxenes (Cawood 1983) or Ca-amphiboles (Morton 1984b) permitted the reconstruction of plate tectonic setting in both cases. Mange-Rajetzky & Oberhänsli (1982) deduced the timing of uplift and erosion of subduction complexes from studies of different generations of blue sodic amphiboles in the Peri-Alpine Molasse. Morton (1985b, 1987) used garnet chemistry to trace provenance.

For rapid gathering of qualitative information, grains may be mounted on sticky tape or partially embedded on a microscope slide in adhesive. If sufficient grains are available, then they can be mounted in epoxy resin (see description on p. 17).

Mounting a few, small important grains without damaging or losing them is often a problem. Jürg Meggert, in the Mineralogical Institute of the University of Berne, obtained excellent results using moulds filled with Buehler No. 20–3400 AB Transoptic powder. Moulds are made in a Buehler 'Simplimet' Hydraulic Press (obtainable from Buehler Ltd, Metallurgical Apparatus, 2120 Greenwood St, Evanston, Illinois 60204, USA and Buehler-Met GmbH, Lessingstrasse 66/68, D-46 Dortmund, West Germany). About 7 g of Transoptic powder will produce 25 mm diameter and 10 mm thick discs. The grains are placed onto a removable mould base, covered immediately with some powder, and then transferred into the press. The rest of the powder is added and the pressure and temperature are set. The mould is formed within 30 minutes and is left to cool completely under pressure in the machine. The melt surrounds the grains, but they will not move and they remain in place very near to the surface of the disc. The disc is then polished with progressively finer diamond pastes and finished by hand on a polishing cloth.

The scanning electron microscope

The scanning electron microscope (SEM) is probably the most frequently used auxiliary instrument in heavy mineral studies. A modern SEM is equipped with an energy-dispersive X-ray spectrometer which facilitates elemental analysis and thus assists mineral identification. Robson (1982) described a computerized SEM and energy-dispersive spectroscopy (EDS) technique used to identify and count the constituents, including heavy minerals, of unconsolidated marine sediments.

For heavy minerals, the SEM is used to inspect general morphological characteristics (Robson 1984, Mallik 1986). However, the majority of studies focus on the examination of surface textures. These, revealed by the SEM in great detail, mirror the effects of subaerial or subsurface dissolution processes and aid assessment of post-depositional diagnetic modifications (Setlow & Karpovich 1972, Hemingway & Tamar Agha 1975, Setlow 1978, Morton 1979a, 1984b, Berner & Schott 1982, Maurer 1982). Figure 3.4 shows some common dissolution features on garnet, pyroxene, amphibole and staurolite.

Surface textures of quartz grains have been used for over a decade (Krinsley & Doornkamp 1973, LeRibault

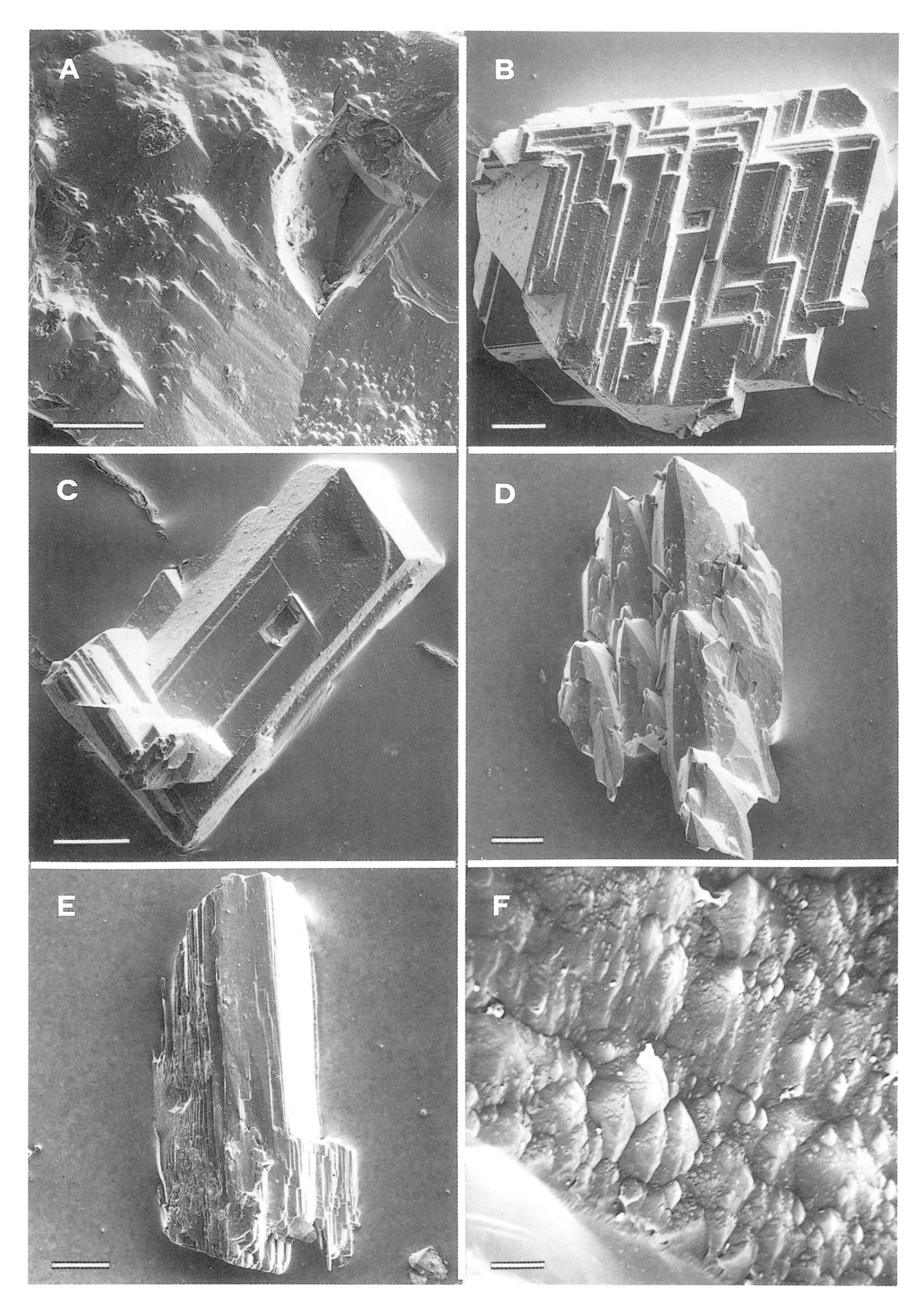

Figure 3.4 Scanning electron photomicrographs illustrating prominent etch features
(A) Garnet showing mamillae and incipient etch facets (Devonian, North Sea); (B) garnet with well developed etch facets (Triassic, North Sea); (C) garnet grain showing advanced stage of etching (Triassic, North Sea); (D) augite dominated by 'hacksaw' terminations (Oligocene, Barrême Basin, France); (E) Ca-amphibole (magnesio-hastingsitic hornblende) displaying a combination of smooth crystal faces (produced during sample preparation) and etching along prominent cleavages (Oligocene, Barrême Basin, France); (F) close up of a mamillated surface of a staurolite grain (Oligocene Molasse, Murgenthal, Switzerland). Scale bar represents 30 μm in (A–E) and 5 μm in (F).

1977). Heavy mineral surface textures have received far less attention (see Stieglitz 1969, Setlow & Karpovich 1972, Lin *et al.* 1974, Gravenor 1979), in spite of the fact that the morphology and surface patterns of first-cycle heavy minerals may in fact provide better information on environmental processes than can be obtained from quartz, which is usually polycyclic.

For SEM inspection the grains are mounted on standard aluminium stubs which are covered with double-sided adhesive tape.

Cathodoluminescence

The property of certain minerals to emit energy in the form of light when bombarded by electrons is known as cathodoluminescence (CL). This property is prompted by the presence of impurities ('activators'), often manganese or rare earth elements incorporated in the mineral lattice; on the other hand it may be inhibited by other elements ('quenchers'), e.g. iron. The relative proportion of quenching and activating impurities will determine the cathodoluminescence properties of the host mineral.

CL is observed either using the electron microprobe, which is useful as it also reveals the identity of the 'activators' (Smith & Stenstrom 1965, Görz *et al.* 1970), with the SEM, or with CL microscopes (Herzog *et al.* 1970). Nuclide Corporation AGV Division (Box 315, Acton, Mass. 01720, USA) has recently introduced an energy-dispersive spectroscopy (EDS) accessory which can be fitted to a Nuclide Luminoscope. This facilitates elemental identification as well as qualitative and quantitative analysis, while viewing the specimen. The hot cathode luminoscope constructed by Zinkernagel (1978), modified and improved by Ramseyer (Ramseyer 1983, Matter & Ramseyer 1985), yields excellent results with sandstones and permits observation of faint and quickly fading luminescence.

The capacity of CL for revealing internal textures, the partitioning of trace elements, etc. – properties not detectable through the petrographic microscope – has made it an important tool in sedimentary petrology (Nickel 1978).

Heavy minerals have so far received little study under CL. Only the luminescence of apatite was studied systematically (see below). Reports on the CL behaviour of other heavy minerals merely describe the luminescence colour and intensity of a particular species, often without referring to the host rock from which it was extracted. Dudley (1976) and Mariano (1977) reported the luminescence of a few detrital heavy mineral species from sediments. Data available to date on heavy minerals are shown in Table 3.1. Additional information can be found in Marshall (1988, pp. 37–56).

Smith & Stenstrom (1965) recognized that the luminescence colours of apatite show affinities to particular parent rock types: e.g. alkaline plutonic rocks from Greenland yielded a lavender-coloured luminescence; Precambrian microcline perthites from Finland contained minute yellowish-luminescing crystals, whereas those of pegmatites exhibited a variety of colours, including green. Some greywackes studied by these authors contained two or three types of apatites, each recognizable by their individual luminescence properties.

In addition to manganese, Portnov & Gorobets (1969) identified additional activators in apatite as Ce^{3+}, Sm^{3+}, Eu^{2+} and Dy^{3+} ions. They studied the correlation between activators and luminescence colours and established four different groups, each one associated with particular petrogeneses (after Nickel 1978):

- I Bluish-violet CL: apatites from carbonatites, and from mafic to ultramafic rocks.
- II Lilac CL: apatites from alkalic rocks.
- III Pinkish orange to yellow CL: apatites from granitoids, and from their genetically related mineralizations.
- IV Yellow CL: apatites from metasomatic phlogopite deposits in metamorphic rocks.

Mariano & Ring (1975) examined the luminescence of both apatites from carbonatites and genetically associated igneous rocks (the apatites displayed light blue to violet luminescence which was activated by the presence of europium) and apatites from low-temperature porphyry copper deposits (exhibiting bright yellow luminescence which was activated by Mn^{2+}). They found that the degree of europium activation in apatites is specific for different petrogenetic environments. It may be strongly dependent on crystallization temperature: for example low-temperature crystallization conditions inhibit lanthanide intake in apatite. Under these conditions Mn^{2+} is the likely activating cation.

These observations have significant implications. The tendency of apatite to select particular activators under different crystallization conditions makes it a valuable indicator of specific source-rock types.

The differing luminescence colours and internal textures of zircon may also prove helpful in defining certain parentages. Observations by K. Ramseyer and MM on hyacinth zircons from West African alluvial concentrates, on zircons from the Oligocene–Miocene Numidian Flysch Formation of northern Tunisia, and from orthogneisses from the southern Alps, indicate their potential use. The West African and orthogneiss zircons exhibited blue luminescence and extensive internal zoning parallel to their long axis (Figure 3.5A and B, page 143) which was invisible under the polarizing microscope. The Numidian Flysch contained two varieties of zircons, one type showing non-luminescent overgrowth on a yellow luminescent core (Figure 3.5C, page 143) and the other displaying blue luminescence and sharply defined internal zoning, not detected under the petrographic microscope. These observations bear significant implications regarding the CL properties of zircon and the corresponding petrogenesis which may be deduced by examining zircons under CL.

The CL properties of zircon and apatite, ubiquitous species in both ancient and young sediments, have much to offer in clastic sedimentary petrology. Sandstone units may be characterized by the presence of zircons with distinctive luminescence or internal texture (or both). These features may also serve to correlate or distinguish between certain sandstone units or formations. The homogeneity or variability of apatite and zircon luminescence colours within a sandstone unit can also be valuable in tracing parentage, or in indicating specific or mixed provenance. (However, these properties have yet to be exploited.)

Uncovered and polished grain mounts are often necessary for high-quality CL analyses. For the hot cathode luminoscope the most suitable mounting media are the high-temperature-resistant resins. A very thin coating of carbon, gold or aluminium will protect the slides from destruction by heat.

Table 3.1 Cathodoluminescence of heavy minerals. Compiled from data in the literature.

Mineral	Locality	Luminescence colour	Activator	Reference	Nonluminescent Minerals
Amphibole	Grenville Rocks, Talcville N.Y. USA	dark red and greyish-yellow	Calcium-poor and manganese-rich	Weiblen, 1965	
Andalusite	Hill of Cabrach, Aberdeensh., Scotland	dark blue to light blue	Ti, Cr	Chinner et al. 1969	Enigmatite, Epidote, Garnet, Glauconite, Rutile, Staurolite, Tourmaline (Dudley, 1976)
	Beach sand	brilliant red	Fe^{3+} substituting for Al^{3+}	Mariano, 1977	
Apatite	see text for details	yellow, violet or blue, green	Mn^{2+} and REE Eu^{2+}, Eu^{3+}, Sm^{3+}, Dy^{3+}, Tb^{3+} colour depends on REE present and proportions	see text for details, also Mariano, 1977	Monazite (Mariano, 1977)
Augite	Perry Formation, New Brunswick, Canada & Maine USA	greenish-yellow	n.i	Schluger, 1976	Enstatite, Bronzite, Hypersthene (Reid et al.1964)
Biotite	n.i	blue	n.i	Görz et al. 1970	
Cassiterite	n.i	blue	n.i	Görz et al. 1970	Almandine, Andalusite, Axinite, Sphene, Topaz (Görz et al. 1970)
	Beach sand	intrinsic	-	Mariano, 1977	
	Hydrothermal	yellow, zoned	Fe, Ti	Hall & Ribbe, 1971	
	Hydrothermal	blue, zoned	W	Hall & Ribbe, 1971	
Corundum	n.i	blue	n.i	Görz et al. 1970	
	Beach sand	red	Cr^{3+}	Mariano, 1977	
Diopside	n.i	bluish red, patchy	n.i	Görz et al. 1970	
Epidote	Porphyry basalt, Oslo region	green, irregular	n.i	Smith&Stenstrom,1965	
Fluorite	Porphyry copper deposit,hydrothermal	rich blue	Eu^{2+}	Mariano & Ring,1975	
Forsterite	n.i	reddish patchy	n.i	Görz et al. 1970	
Garnet	n.i	dull yellow	n.i	Schluger, 1976	
	Beach sand	red, yellow	Cr^{3+}, Mn^{2+}	Mariano, 1977	
Grunerite	n.i	reddish blue, decay to blue	n.i	Görz et al. 1970	

Jadeite	n.i	wine-red, rapid decay	n.i	Görz et al. 1970	
Kunzite	n.i	bright pink	Mn^{2+}?	Görz et al. 1970	
Kyanite	n.i	lamellar, var.of bright red, purple, deep red, patchy	n.i	Chinner et al.1969	
	Soil sample	bright to dull pink, pale blue	n.i	Dudley, 1976	
	Beach sand	brilliant red	Fe^{3+} substituting for Al^{3+}	Mariano, 1977	
Mica aggregates	n.i	dull purple	n.i	Schluger 1976	
Monazite	Soil sample	dull reddish-brown	n.i	Dudley, 1976	
Muscovite	n.i	wine red to black	n.i	Görz et al. 1970	
Pyrope	n.i	no Cl, or very weak, patchy	n.i	Görz et al. 1970	
Scheelite	Porphyry copper deposit,hydrothermal	intrinsic +	Sm^{3+},Eu^{3+},Tb^{3+},Dy^{3+}	Mariano & Ring 1975	
	Beach sand			Mariano, 1977	
Sericite-chlorite	n.i	greenish-purple	n.i	Schluger, 1976	
Sillimanite	n.i	dull blue	n.i	Nickel, 1978	
	Beach sand	brilliant red	Fe^{3+} substituting for Al^{3+}	Mariano, 1977	
Talc	New York	white	n.i	Reid et al. 1964	
Topaz	Soil sample	bright light blue	n.i	Dudley 1976	
Tourmaline	Perry Formation, New Brunswick, Canada & Maine USA	dull pink bright green or dull yellow	n.i	Schluger, 1976	
Tremolite	Grenville Rocks, Talcville N.Y. USA	golden yellow	n.i.	Weiblen, 1965	
Wollastonite	n.i	bright green	Mn^{2+}	Long & Agrell, 1965	
Zircon	n.i	yellow and blue zoned	n.i	Görz et al. 1970	
	Perry Formation, New Brunswick, Canada & Maine USA	yellow or green	n.i	Schluger, 1976	
	Beach sand	yellow	intrinsic and Dy^{3+}	Mariano, 1977	
n.i - not indicated					

4 Presentation and numerical analysis of heavy mineral data

After analysis, data are tabulated and the quantities of the individual heavy minerals are expressed in relative or weight percentages. This gives the first insight into the heavy mineral spectrum and provides a basis for deciding on further data treatment.

Cumulative heavy mineral compositions can be shown graphically when the heavy mineral variations are plotted horizontally against distance or vertically against depth. Pie-diagrams are informative in depicting the spatial distribution of the heavy minerals being studied and are useful in highlighting differences in heavy mineral compositions between sediments from different petrological provinces in a study area. Mineral dispersal patterns are best illustrated with isopleth maps. Binary plots (scatter diagrams) and ternary (triangular) diagrams permit characterization of a group of samples, particularly when data points cluster within a certain field on the diagram. Such characterization in turn helps to outline a unit, a petrological province or formation.

Although visual inspection of figures allows detection of the more obvious trends, the less marked differences can easily be overlooked, and hence the employment of advanced numerical techniques is of considerable importance if data assessment is to be objective. Modern computer techniques allow selection of the most suitable programs from a wide range of available options.

Imbrie & van Andel (1964) introduced computer analysis for the interpretation of heavy mineral data, using *Q-mode factor analysis*. This analysis is most applicable in studies involving complex and remote source regions and far travelled, mixed sands. The Q-mode analysis produces maps showing regional variability patterns useful in revealing the areal distribution of mineral assemblages. This method also assists the pinpointing of principal sediment supply areas and the correlation of distribution patterns with geological events and processes.

Q-mode factor analysis, coupled with varimax rotation, for processing heavy mineral data has yielded useful results in several Recent depositional basins such as the Orinoco–Guyana Shelf (Imbrie & van Andel 1964), the Bering Sea (Knebel & Creager 1974, Gardner *et al.* 1980), the South Texas Outer Continental Shelf in the Gulf of Mexico (Flores & Shideler 1978), the Bristol Channel, UK (Barrie 1980) and the East China Sea (Lirong *et al.* 1984). Galehouse (1967) applied Q-mode factor analysis to determine the degree of mixing and the compositions of source assemblages for the Pliocene continental Paso Robles Formation, California. Maurer *et al.* (1978) and Maurer & Nabholz (1980) used Q-mode factor analysis of heavy mineral data to distinguish several depositional units within the Oligocene–Miocene Molasse sequences of the Linden-1 and Romanens-1 boreholes from Switzerland.

Stattegger (1987) applied the extended Q-mode factor analysis of Klovan & Miesch (1976) for a statistical treatment of heavy mineral data in order to investigate its potential for reconstructing source-rock lithologies and provenance. This method was first tested on heavy minerals from river sediments of known provenance from northern Austria. End members, calculated by the extended Q-mode factor analysis, led to relevant source rock lithologies and, in addition, they yielded information on mixing effects. Subsequently Stattegger used the model to treat heavy mineral assemblages of the Alpine synorogenic Gosau Formation (Late Cretaceous). The results permitted the differentiation of three major lithologies, and indicated the particular plate tectonic setting of the depositional areas and that of the hinterland.

Principal component analysis – one of the multivariate techniques and contributing part of all factor analytic schemes – transforms the original variables of a data matrix to a set of theoretical independent variables called 'principal components' (Rice *et al.* 1976, Davis 1980) and deals with eigenvectors of a valence–covalence matrix. These components can be used to explain most of the variance within a much larger set of original variables. Rice *et al.* (1976) used principal component analysis to identify major heavy mineral distribution patterns along the southern California coast. Pirkle *et al.* (1984) employed the same technique to elucidate processes controlling the accumulation of Recent heavy mineral-rich sand deposits of north-eastern Florida. In a subsequent study Pirkle *et al.* (1985) successfully unravelled the original heavy mineral composition and provenance of two economically important formations from the Miocene of Florida. This approach also facilitated the identification of two different sedimentary facies within the large Miocene delta.

Linear discriminant analysis was employed by

Demina (1970) to differentiate between two contrasting sediment derivation paths and by Gwyn & Dreimanis (1979), who recognized discrete heavy mineral suites useful in distinguishing between different glacial lobes. Stattegger (1982) succeeded in delineating Ordovician and Upper Palaeozoic rocks in Steiermark, Austria, by use of heavy mineral data treated by the above technique.

Flores & Shideler (1982) employed *stepwise multiple discriminant analysis* in a study of the Outer Banks barrier system of North Carolina, hoping to distinguish between foreshore, berm and dune sub-environments, but this approach was unsuccessful.

Cluster analysis of heavy mineral data was tried by, among others, Rice *et al.* (1976), Maurer *et al.* (1978) and Maurer & Nabholz (1980). Cluster analysis is generally used combined with another multivariate routine, usually factor analysis.

The studies mentioned above all give more details on the precise nature of the mathematical modelling and computing techniques employed in each case. The reader is referred to these works for further information.

5 Application of heavy minerals

If a man will begin with certainties,
he shall end in doubts; but if he will be content to begin with
doubts, he shall end in certainties.
Francis Bacon (1561–1626)

The following examples demonstrate the versatility and wide application of heavy mineral techniques.

The properties of heavy minerals allow an insight into the petrological character of sediment source terrains and permit their extensive use in *tracing provenance*. An association of characteristic heavy minerals is intimately related to particular source lithologies and may often be correlated with an identifiable source terrain. Recognition and mapping of the spatial distribution of heavy mineral associations furnishes information on sediment *dispersal patterns*. A survey of the regional variation of heavy mineral compositions enables delineation of *heavy mineral provinces*. A heavy mineral province is characterized by the presence of volumetrically important heavy mineral species that are absent or rare in adjacent provinces. Analysis of heavy mineral content through a vertical profile (e.g. in outcrops or in boreholes) may also reveal the presence of heavy mineral zones, typified by volumetrically important heavy mineral species which are absent or rare in the underlying or overlying beds.

Heavy mineral studies on *modern basins* have provided actualistic analogues which lead to a better understanding of the geological history of their ancient counterparts. Baak (1936) demonstrated the co-relationship of source areas, transport processes (fluvial, glacial and marine) and the five heavy mineral provinces in the surficial sediments of the southern North Sea.

The Gulf of Mexico and adjacent regions have been a fruitful area for heavy mineral studies (Goldstein 1942, van Andel 1960, van Andel & Poole 1960, Davis & Moore 1970, Flores & Shideler 1978). These studies revealed the presence of important heavy mineral provinces and provided clues to the pathways of the Pleistocene fluvial network on the shelf during low sea-level stands. Heavy minerals also permitted mapping of Pleistocene river mouths and indicated the extent of reworking and redistribution of the shelf sediments during the Holocene transgression.

Other studies employing heavy minerals to yield information on sedimentary processes include those of McMaster & Garrison (1966) in a study of the South New England shelf, Hubert & Neal (1967) working in the North Atlantic, Kelling *et al.* (1975) working offshore from the Mid-Atlantic USA, Knebel & Creager (1974), Gardner *et al.* (1980) in studies of the Bering Sea, Gazzi *et al.* (1973), Rizzini (1974) working in the northwestern Adriatic Sea, Belfiore (1981) in the western Mediterranean, van Andel (1964) in the Gulf of California, Barrie (1980) in a study of the Bristol Channel UK and Mange-Rajetzky (1983) in the eastern Mediterranean.

Lateral movement of beach sand was detected using heavy mineral studies in Portland, Australia (Baker 1956), on the Rhode Island shore (McMaster 1960), on the continental shelf off Accra, West Africa (Brückner & Morgan 1964), and on the northern coast of Columbia (von Erffa 1973).

The study of van Andel (1955) on the sediments of the Rhône delta is an example of how heavy minerals may be used to indicate sediment distribution patterns within a single river delta system.

Mange-Rajetzky (1979) studied the heavy mineral composition of Quaternary–Recent sediments along the southern Turkish coast with the view to contributing to the modelling of molasse-type sedimentation. The analyses revealed distinct heavy mineral associations which were linked to specific source-rock compositions. Sedimentary petrological provinces, delineated on the basis of light and heavy mineral data, mirrored their relationship to prominent source-rock lithologies and also indicated processes distributing the sediments derived from them.

Heavy minerals are often the only means of reconstructing provenance in sedimentary sequences. A comprehensive study by Füchtbauer on the *Molasse of the Northern Alpine Molasse Basin* (1964, 1967; summarized in English in 1974) has contributed greatly to the understanding of the type basin of Molasse sedimentation. Heavy mineral data provided essential clues for the interpretation of sediment provenance, assisted in distinguishing between the loads of the major alluvial fans feeding the basin, and highlighted sediment transport and dispersal directions within the basin. Allen & Mange-Rajetzky (1982), in a study of sediment dispersal and palaeohydraulics of Oligocene rivers in the eastern Ebro Basin, combined heavy and light mineral analyses with palaeodischarges estimated for rivers of differing parentage. The integration of these two approaches resulted in the delineation of two major sedimentary petrological provinces and several sub-provinces, all of which showed intimate links with two opposing source areas: the Pyrenees to the north and Catalanides to the south-east. Prolonged tectonic activity in the Pyrenean chain promoted an increasing Pyrenean dominance through time and this is reflected in the progressively greater areal extent of the Pyrenean province. Heavy mineral analyses assisted in revealing the sources of sand supply and helped to elucidate mechanisms governing the accumulation of tidal sandbanks in the Burdigalian sea of the Swiss Molasse Basin (Allen *et al.* 1985). A combined study of rock fragments and heavy minerals in the Miocene Marnoso Arenacea Turbidite Formation in the northern Apennines (Gandolfi *et al.* 1983) shed light on the nature and evolution of this complex deep-sea fan system. Six sediment source areas were identified, dispersal patterns were mapped and the progressive north-eastward migration of source terrains and depocentres during the Apenninic orogenesis was recognized. Hansley (1986) applied the stratigraphic variation of heavy minerals and some of their varietal types in a study of the Morrison Formation, New Mexico. This work permitted the determination of provenance and the reconstruction of diagenetic episodes of this economically important uranium-bearing formation.

Use of mineral chemistry and varietal studies can be readily applied in reconstructing provenance. A potential approach is outlined on pages 18–19.

It is commonly believed that heavy minerals have a limited value for *time-stratigraphic correlation*. Results have been unsatisfactory in regions where the source areas (usually cratonic terrains) and sediment accumulation rate remained constant for a considerable length of geological time. Carroll (1940) stressed that a large number of well located samples is necessary for the accurate correlation of subsurface sediments with their outcrop counterparts.

Dense sampling in closely spaced boreholes yielded accurate stratigraphic correlation of heavy mineral zones in the San Joaquim Valley, California (Reed & Bailey 1927). Similarly, identification and correlation of stratigraphic units by means of heavy minerals have proved successful in Venezuela (Feo-Codecido 1956). Koen (1955) applied heavy mineral studies in the correlation of members within the Karroo System in the northern part of South Africa. His work permitted local correlation and allowed clarification and correlation of the existing stratigraphy.

Lithostratigraphic units are commonly associated with particular heavy mineral assemblages. Depositional breaks or unconformities are often characterized by marked changes in the heavy mineral suites. This emphasizes the *stratigraphic significance of heavy minerals*. Tieh (1973) applied heavy mineral analyses and was able to characterize and distinguish stratigraphic horizons of the Eocene to Miocene sediments of central California. Stattegger (1976) showed that each formation of the Cambro-Ordovician to post-Variscan (Stephanian) clastic series in the axial zone of the Pyrenees is identifiable by a characteristic heavy mineral suite. Time-stratigraphic units of the Peri-Alpine Molasse can be clearly recognized by their distinctive heavy mineral compositions (Mange-Rajetzky & Oberhänsli 1982, Maurer 1983). Woletz (1963, 1967) analysed Cretaceous and Palaeogene sediments from the Austroalpine domain and found that several tectonic and stratigraphic units can be typified by their particular heavy mineral associations. These units indicated links to parentages with which they formed intimate tectonic relationships at the time of sedimentation. Weissbrod & Nachmias (1986) used heavy mineral data to define stratigraphic units within the Precambrian to Mesozoic 'Nubian Sandstone' sequence of southern Israel, southern Jordan and Sinai. These units coincided with previously defined lithostratigraphic units and could be used for stratigraphic correlation. Changes in the heavy mineral assemblages aided the recognition of paraconformities that were otherwise not readily apparent.

Briggs (1965) developed statistical means and modal separation for the recognition and correlation of independent mineral provinces.

Heavy mineral studies have been proved to be important contributors to the analysis of *sedimentation associated with tectonically active hinterlands*. The vertical and stratigraphic evolution of heavy mineral assemblages of the basin fill provide a mirror of tectonic events and the erosion history of the source terrains. In sediments, eroded from orogenic belts, the heavy mineral suites commonly become increasingly complex in a vertical sequence, this evolution reflecting the uplift and erosion of successively more complex lithologies in the source area (unroofing). The sequence of heavy mineral zones are seen in the inverse order of the sequence at source.

Heavy mineral assemblages are generally well preserved in sedimentary settings associated with pronounced topographic relief because of the consequent high sedimentation and burial rates. The Lower

Oligocene to Upper Miocene Peri-Alpine Molasse basin evolved in response to tectonic events in the Alpine mountain belt. Consequently, the succession of heavy minerals in the Molasse sediments mirror the progressive uplift of Alpine structural units and thrust movements. Minerals diagnostic of early Alpine metamorphism are well preserved in these sediments, even though in the source area they may have been overprinted by later metamorphic episodes, or either eroded or buried under thrust sheets in the hinterland. Analysis of these mineral assemblages provides much useful information, facilitating the more accurate reconstruction of the orogenic and sedimentary histories of the region. The first appearance of index minerals from each of the multiple phases of Alpine metamorphism in the sediments indicates the time when erosion of their respective parent rocks commenced (Vatan 1949, Füchtbauer 1974, Mange-Rajetzky & Oberhänsli 1982, 1986, Maurer 1983). The frequently observed marked changes in heavy mineral assemblages at prominent stratigraphic boundaries within the Molasse sequence document important changes, as well as reflecting evolution in sedimentary environments and changing transport mechanism through time.

Winkler & Bernoulli (1986) reported the occurrence of detrital high-pressure–low-temperature index minerals in a late Turonian flysch sequence of the eastern Alps. These minerals were formed by Mid-Cretaceous subduction-related metamorphism, and their presence in the late Turonian sediments indicates uplift of their parent rocks from about 20 km depth within a rather short time interval. This has important tectonic implications bearing on the possible mechanism responsible for such rapid uplift.

Heavy mineral studies have been widely used to *clarify specific problems* and to *complement other types of research*. The presence and abundance of euhedral zircon indicated volcanism in the Lower Tertiary of central Texas (Callender & Folk 1958). Weaver (1963), Spears (1982), and Winkler *et al.* (1985) emphasized the interpretative value of heavy minerals in recognizing volcanic clays and distinguishing between primary and secondary bentonites.

Heavy minerals of air-fall origin in a detrital assemblage reflect contemporary volcanism, and their chemistry may provide clues to the nature of the volcanism (Morton 1982a). Heavy mineral assemblages have also been used in *mapping and correlating tephra layers* (Juvignè & Shipley 1983).

Investigating the mineralogy of glacial sands can prove a valuable means of *bedrock mapping in glaciated terrains* (Callahan 1980). Gwyn & Dreimanis (1979) investigated the heavy minerals of tills in the Great Lakes region of North America, and succeeded thereby not only in distinguishing various individual glacial lobes but also in determining their respective provenance and bedrock lithology in each case. Heavy minerals incorporated in ice-rafted debris may be of use in pinpointing the source regions of these materials (Alam & Piper 1981), especially when applied in conjunction with other petrographic analyses. Heavy minerals of stream sediments in *remote terrains* can yield valuable information on rock formations in their scarcely known or inaccessible catchment areas (Mange-Rajetzky 1981). An interesting application of heavy minerals in the detection of the sources of Recent *dust fallout deposits* caused by dust-storms in Kuwait was reported by Khalaf *et al.* (1985).

In recent years *geochemical methods* have been increasingly used in the study of heavy mineral assemblages. Silver & Williams (1981, 1982) investigated the morphology and physical properties of zircon populations, together with their chemical and isotopic compositions to *assist the modelling of uranium mineralization* processes in the Morrison Formation (southern Colorado Plateau and San Juan Basin, New Mexico). Their approach can also be applied to other heavy mineral species in detrital sediments, and may help to obtain more refined information on the nature of provenance.

Thomas *et al.* (1984) used analysis of rare earth elements (REE) incorporated in sphene, allanite and monazite, to distinguish particular source rock characteristics. They demonstrated the potential of their method in distinguishing between sediments derived from different sources and proposed its wider use for the correlation of unconsolidated sediments. Owen (1987) demonstrated that concentration of hafnium in detrital zircons is a viable indicator in provenance studies. Owen used this method to reveal petrogenetic relationships between sandstones which were derived from the same proximal source.

Heavy minerals of alluvial deposits have long been used for *prospecting* for gold, platinum and diamonds (e.g. Raeburn & Milner 1927). Zeschke (1961) used the presence of scheelite in stream sediments to trace valuable tungsten deposits in Pakistan. Goldfarb (1981) located mineralized skarns in the southern Sierra Nevada, California, by applying geochemical techniques to heavy minerals extracted from river sediments down stream.

Mertie (1979) demonstrated the value of heavy minerals in *regional geological exploration*. Saprolitic accessory minerals assisted the delineation of the economically significant monazite belt in the southeastern USA. Accessory minerals also proved useful in indicating the metasedimentary or migmatic origin of granites of this region.

The analysis of heavy mineral concentrates from alluvial and colluvial deposits as well as from soils is a technique widely used in *kimberlite prospecting* (Kresten *et al.* 1975, Leighton & McCallum 1979, Hearn & McGee 1983). Brown (or chromian) spinel can be a valuable signal of *diamond placers*. In Namibia it is commonly associated with Recent diamond placers (Zimmerle 1984) and it is found in fossil diamond placers in the northern Urals (Bekker *et al.* 1970).

Several earlier papers on the economic analysis of heavy minerals in sediments were compiled and briefly reviewed by Luepke (1985).

The contribution of heavy mineral studies to the solution of problems in basin analysis, petroleum exploration and reservoir management is considerable. When allied with other sedimentological and/or stratigraphical and geochemical techniques, they have much to offer towards a better understanding of the history of a basin. The study of heavy minerals proves especially informative in lithologically uniform clastic hydrocarbon reservoirs, as they can reveal the petrological heterogeneity of the successions. Such heterogeneity is highlighted by the presence of distinct heavy mineral zones in reservoir successions, recognized in the distributional patterns of heavy mineral species and paralleled by similar trends in the ratios of selected mineral pairs. Major changes in the heavy mineral spectrum usually coincide with peaks in the plot of the ratios, thus permitting delineation of independent heavy mineral zones. Heavy mineral zones with similar characteristics and containing similar varietal and morphological heavy mineral types can be correlated. Therefore the use of heavy minerals has considerable importance in basin-wide lithostratigraphic correlation.

Heavy minerals are sensitive indicators of sedimentary processes and are useful in signalling changes of the depositional environment or facies. This capacity can be used to assist facies analysis, important in evaluating reservoir properties. The morphology of many species may help to distinguish and correlate aeolian horizons in successions where fluvial and fluvio-lacustrine sediments are interbedded with aeolian deposits (Allen & Mange-Rajetzky 1989 and in preparation). Recording characteristic varietal types and computing their ratios provide important clues to the identification and comparison of different sand bodies and to unravelling the recycling paths of older sediments into successively younger formations. The presence or absence of certain varietal types of the ultrastable suite is especially important in this respect.

Structural events in the hinterland, resulting in the exposure of new lithologies, are mirrored by distinct changes in the heavy mineral compositions. These changes are recognized as marker horizons in the heavy mineral spectra and can be correlated basin-wide. Heavy mineral assemblages are useful indicators of proximality/distality of a location to the source regions; they provide clues to the definition of sediment dispersal patterns, thereby helping the assessment of sand body geometry.

In the knowledge of their chemical stability, heavy minerals contribute to the evaluation of diagenetic episodes and burial history of reservoir successions. Heavy mineral analyses can be performed on drill-cutting samples, thus providing valuable data for the interpretation of uncored sequences. Studies by Morton on the North Sea (1979b, 1982b, 1984b, 1985b, Hurst & Morton 1988) and my published (Allen & Mange-Rajetzky 1989, Mange-Rajetzky 1989) and unpublished data on petroleum reservoirs in different countries have assisted in the reconstruction of the depositional environments of marine and terrigenous clastic sediments, and have helped to provide an extensive information data base of use in more accurate modelling of reservoir development and quality distribution.

Although the purpose of this book is to provide a guide for the identification and use of transparent heavy minerals, it is necessary to mention the importance of the **opaque suite**, which often constitutes the bulk of the heavy mineral fraction.

Use of polished sections and reflected-light microscopy are essential for the identification of opaque grains, but this technique is generally neglected by sedimentary petrologists. However, the value of opaques in provenance reconstructions has been demonstrated in the few studies carried out (Blatt 1967, p. 1038). Sanders & Kravitz (1964), by summarizing the results of several studies, emphasized the importance of opaque mineral determinations in palaeomagnetic research, economic geology and in the understanding the origin of red beds.

Stumpfl (1958) pointed out the significance of many opaque species in interpreting hinterland lithologies. An assemblage of ilmenite containing finely exsolved lamellae, together with 'pure' magnetite and ilmenite–haematite, is typical of acid plutonic source terrains. Ilmenite, and ilmenite with exsolved broad titano-magnetite lamellae, occur in basic plutonics. Pseudobrookite is characteristic of volcanic rocks. Hilmy *et al.* (1971) investigated the opaque constituents of recent beach sands in Kuwait. Their analysis has revealed a diversity of opaque minerals and has proved useful in tracing the source of these beach sands to various rock types present in Turkey, Jordan, Syria and Iran. Luepke (1980) used the magnetic opaque fraction of beach sands on the south-western Oregon coast to distinguish between source and wave-sorting effects on the distribution of heavy minerals in the area.

Darby and Tsang (1987) focussed on the variation in elemental composition of detrital ilmenite grains in the sediments of three large drainage basins in Virginia, USA. Their analyses indicated that ilmenite compositions remained virtually unchanged along major river courses and also that the compositions were specific for different drainage basins within their study area. Granitic, mafic igneous, metamorphic and sedimentary sources were clearly identifiable from different ilmenite compositions. Darby & Tsang concluded that this approach can provide an additional means for distinguishing between the sediments of adjacent drainage basins and for determining sand provenance.

PART II

DESCRIPTIONS OF HEAVY MINERALS

6 Introduction

Go my sons, buy stout shoes, climb the mountains, search the valleys,
the deserts, the sea shores and the deep recesses of the earth.
Mark well the various kinds of minerals,
note their properties and their mode of origin.

Petrus Severinus (1571)

It has been said that 'a picture is worth a thousand words', but a picture accompanied by a descriptive analysis is even more valuable. We hope that the full-colour photomicrographs in this part of the book will assist the accurate identification of heavy minerals by complementing the descriptions of grain morphology, optical and physical properties.

Multiple grain mounts of each heavy mineral species were prepared for the images. Heavy minerals were selected from a variety of sedimentary environments in order to provide the most informative illustrations. Grains affected by dissolution and the most common varietal types are also included.

Cargille Meltmount series resins were used for mounting the grains and the particular mounting media in each case is indicated at the end of the mineral descriptions. Meltmount 1.539 and 1.582 were employed for minerals with low to medium RI, and Meltmount 1.662 was used for those minerals with a high relief. This approach may present some problems when comparing reliefs, but was necessary to achieve the best possible presentation of optical and physical properties.

The individual grains were first stuck onto a gum arabic base, thus securing their position. The resins were then applied. In spite of painstaking efforts, in a few cases a little air became trapped between grains and the gum arabic film. This appears in the illustrations as one or more small bubbles. These may easily be distinguished from surface patterns and inclusions and should not be confusing.

In order to facilitate the best observation of pleochroism and interference colours, the polarizer was rotated instead of the microscope stage.

The age and locality of the deposit from which the minerals were obtained is indicated for all illustrations, except for those where commercial confidentiality still applies.

The optical diagnosis was confirmed by microprobe analysis for minerals which required chemical means for positive identification (e.g. amphiboles, pyroxenes, carbonate and sulphate minerals and rare species).

Chemical formulae, optical and physical properties and occurrence were taken from Deer *et al.* (1962, 1963, 1978, 1982, 1986). These were complemented in some cases with additional data from Milner (1962), Heinrich (1965), Parfenoff *et al.* (1970), Phillips & Griffen (1981), Boenigk (1983) and from many other publications.

The chemical stability of the heavy minerals is discussed in Part I, and in addition the stability and particular sedimentary occurrence of some prominent species is briefly summarized under 'Remarks'.

The descriptions of heavy minerals are arranged in systematic rather than in alphabetical order. This facilitates liaison between the English and German editions.

Abbreviations and symbols

CB Canada balsam
CL cathodoluminescence
Mmt Meltmount
n refractive index
$r>v$ (or $r<v$) the optic axial angle is greater (or less) in red light than in violet light
RI refractive index
2V optic axial angle
x, y, z the crystal axes
α, β, γ least, intermediate and greatest refractive indices
α, β, γ the vibration direction of the slow, intermediate and fast ray
ε refractive index, extraordinary ray
ω refractive index, ordinary ray
δ birefringence
Δ density

Elongation: positive = length slow
negative = length fast

Length of scale bar: 100 μm

7 Heavy mineral descriptions and colour plates

SILICATES

Olivine Group

Forsterite–Fayalite Series
orthorhombic, biaxial (+) (−)

forsterite Mg_2SiO_4		fayalite Fe_2SiO_4	
$n\alpha$	1.635	$n\alpha$	1.827
$n\beta$	1.651	$n\beta$	1.869
$n\gamma$	1.670	$n\gamma$	1.879
δ	0.035	δ	0.052
Δ	3.22	Δ	4.39

Forsterite (Mg-olivine) and fayalite (Fe-olivine) are the end members of a solid solution series and, when chemical analysis has not been carried out to positively identify a particular member, grains in sediments are usually designated as **olivine**.

Form in sediments: Specimens (a) and (b) **forsteritic olivines**. Grains are fractured, angular, irregular or rounded, often attaining a high sphericity (upper). They frequently enclose minute opaque impurities and show tiny bright-rimmed dissolution hollows and various etch patterns on their surface. Forsterites of serpentinized bodies may be attached to or covered by a yellowish or pale green fibrous serpentine. Olivines of extrusive rocks appear in a variety of forms, and are euhedral, prismatic, irregular with conchoidal fractures, rounded or very well rounded. Inclusions are volcanic matrix, spinel, apatite, pyroxene, plagioclase, zircon and ilmenite. Alteration products of olivine minerals are iron oxides, iddingsite, serpentine polymorphs, etc.

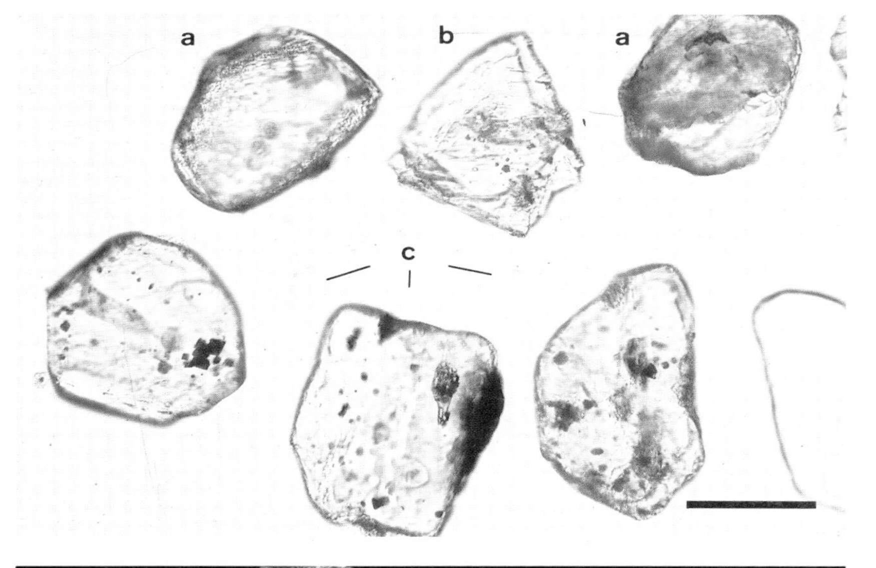

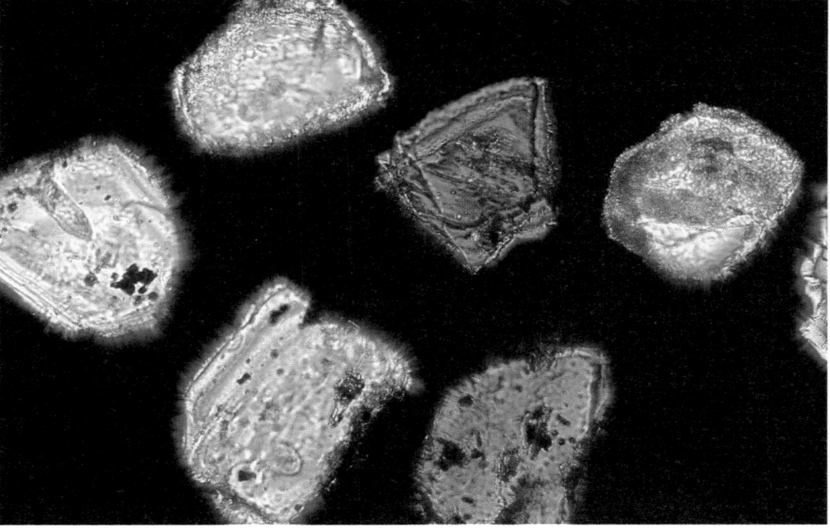

Colour: Forsterite is colourless or very pale green; other olivines are either colourless, honey yellow or pale green.

Pleochroism: Forsteritic olivines are non-pleochroic. Fayalitic varieties display weak yellow to orange-yellow pleochroism.

Birefringence: Moderate to strong birefringence results in vivid interference colours of third-order pink, yellow and green. Several interference colour bands appear arranged either in concentric rings or parallel with the crystal outline. Irregular or patchy polarization colours are not common. Thick grains display very high-order yellowish-white or pink interference tints.

Elongation: Euhedral or elongated grains have parallel extinction.

Interference figure: Forsteritic olivines most commonly yield optic axis figures with an almost straight isogyre, surrounded by numerous bright coloured isochromatic rings. Occasionally bisectrix figures with pale isogyres and large 2V can be observed. Other olivine varieties provide bisectrix figures with either poor or well-defined isogyres and several colour rings. Determination of the optical character is difficult as the dense colour rings obscure observation.

Elongation: Either positive or negative.

Distinguishing features: Olivines are diagnosed by fairly high relief, pale tints or lack of colour and strong birefringence, hence vivid interference colours. The frequency of dissolution features is also distinctive. Serpentinization or iddingsite, when present, is diagnostic. Pyroxenes and amphiboles are usually associated with olivines in sediments and their occurrence may signal its presence. Prismatic olivine grains resemble enstatite, especially when both are serpentinized. However, enstatite shows pyroxene cleavages and often contains exsolution lamellae. The bright interference colours of olivines resemble those of clinozoisite and topaz, but both have better cleavages and lack the characteristic dissolution patterns frequent on olivines. Pale-green olivines can be distinguished from epidote by the higher refractive indices and pleochroism of the latter.

Occurrence: Members of the olivine group are principal constituents of many basic igneous rocks. Forsteritic olivines are important rock-forming minerals of mafics and ultramafics and are also generated by the thermal metamorphism of impure dolomitic limestones. Iron-rich olivines form in alkaline felsic rocks such as some syenites, rhyolites and dolerites. The metamorphism of iron-rich sediments may also generate fayalitic olivines.

Remarks: Due to their highly unstable nature, olivines are usually found only in Quaternary to Recent sediments.

Grains from: (a) Beach sand, southern Turkey; (b) river sand, Centovally, Switzerland; (c) beach sand, Tenerife, Canary Islands (Mmt 1.582).

Zircon

$ZrSiO_4$
tetragonal, uniaxial (+)

$n\omega$	1.923–1.960
$n\varepsilon$	1.961–2.015
δ	0.042–0.065
Δ	4.6–4.7

Form in sediments: Zircon grains appear with a very high relief and they are surrounded by a black halo. Crystals are generally small; their average length in sandstones and arkoses is 0.15–0.25 mm and that in siltstones is 0.05–0.15 mm (Poldervaart 1955). The morphological characteristics of zircon are determined by the physical and chemical conditions during growth; therefore zircon morphology is widely regarded as a petrogenetic indicator. (A condensed review of experimental and petrological studies on zircon is given by Speer 1980.) The elongation (or L/B) ratios of zircons (length divided by breadth) generally indicate the nature of host rocks: in igneous rocks the average elongation ratio is less than 3.0, though there are cases, especially in pyroclastic rocks where it is greater than 3.0 (e.g. some of the grains in upper). Usually the width determines the size of zircon in a particular size fraction, rather than the length. In sediments the morphology of zircon grains varies from sharp euhedral crystals through prismatic and anhedral fragments, grains with gently rounded terminations to well-rounded forms and complete spheres (lower). Rounding is more advanced on larger grains. The surface of the rounded grains may be pitted and frosted. Statistics indicate that zircon of most sandstones is represented predominantly by rounded grains or angular fragments, and euhedral crystals are subordinate. By contrast, zircons of arkoses are mostly sharp euhedral crystals or euhedral grains with rounded terminations, but well-rounded forms are not common. Zircons are often inhomogeneous and several types of zoning have been distinguished on the basis of type and origin. On detrital grains the following types of zoning can be recognized:

(a) Zoning resulting from continuous growth. This is seen as very fine bands generally parallel to the crystal boundary, becoming denser near the periphery (second page middle).
(b) Discontinuous growth is indicated by overgrowth and outgrowth. The former can be total where a zircon grain (euhedral or well-rounded) is epitaxially overgrown by a later generation of zircon and less commonly xenotime, surrounding it on all sides (upper; c). Partial overgrowth develops on one part of the grain (second page middle).

The CL technique is particularly advantageous in revealing the nature of zoning and overgrowth, together with the presence of radioactive and rare earth elements (Sommerauerer 1976, see also pp. 22–23). Outgrowths are pyramidal or hemispherical projections ('saw teeth') growing from a certain face (right, lower). They can be zircon or xenotime. Their delicate nature suggests that they are formed *'in situ'* in sediments, as it is unlikely that these features survive erosion and transportation. Zircons often exhibit a system of fractures. These commonly radiate from the centre, usually starting from inclusions. Hutton (1950) attributed this phenomenon to the radioactivity of the mineral. Inclusions are gas and fluid globules, opaques, zircon, xenotime, monazite, biotite, rutile, cassiterite, tourmaline and opaque dust.

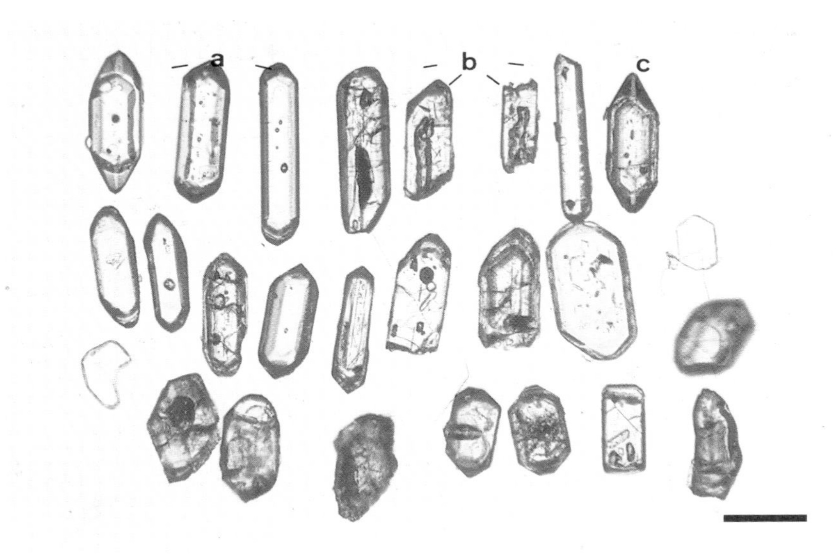

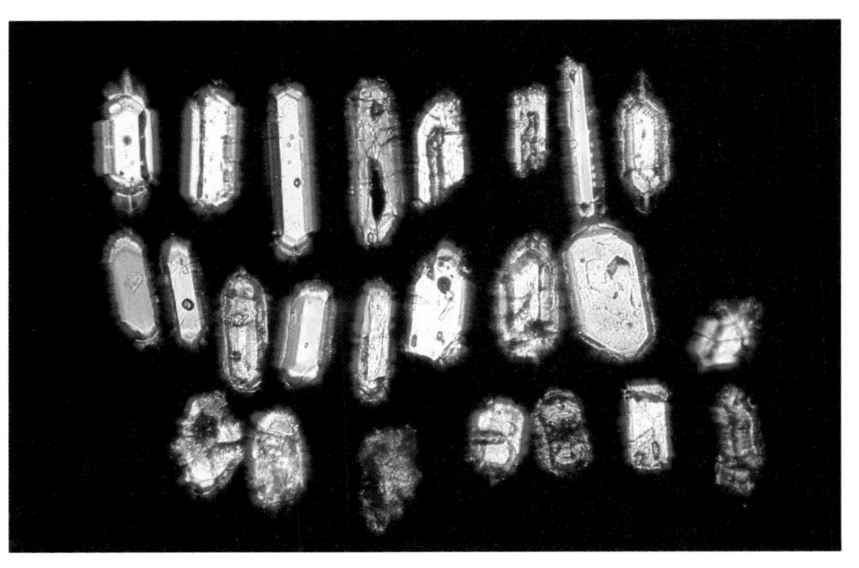

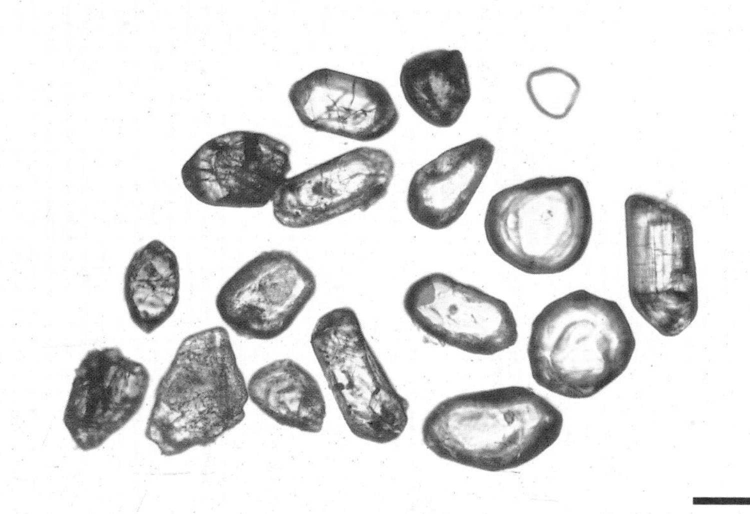

Colour: Zircon is colourless, pink, purple, red, orange, yellow, rarely blue or grey. Hyacinth is the pink–purple variety. The majority of detrital zircons are colourless and pink, purple or yellow shades are subordinate.

Pleochroism: Very weak in stronger coloured varieties.

Birefringence: Zircon has a strong birefringence, hence interference colours are high-order white, whitish-yellow or pink in thicker grains. Thinner crystals and fragments exhibit bright yellow, orange, green and red tints, usually with several interference colour bands.

Extinction: Elongated grains show parallel extinction.

Interference figure: Detrital zircons rarely provide good interference figures. Rounded and prismatic grains commonly yield faint biaxial figures. Grains exhibiting basal sections or rounded grains laying near to their basal pinacoid (recognized by lower birefringence and the tendency of incomplete extinction), provide centred or nearly centred uniaxial figures in a white field, with or without isochromes. Extreme dispersion is seldom visible.

Elongation: Positive, but difficult to observe.

Distinguishing features: Extreme relief, characteristic morphology and strong birefringence, together with the lack of strong colours, are diagnostic of zircon. It is one of the readily identifiable detrital minerals and only in rare cases can it be confused with such grains as sphene, monazite, rutile, cassiterite and xenotime. The birefringence of sphene is stronger, and monazite has a lower relief, yellowish or greenish colours and frequent surface pitting or brownish stains due to decomposition. In addition the characteristic absorption lines of monazite (see under monazite or xenotime) aid a positive distinction. Rutile has a higher relief, deeper colours and is usually pleochroic. Cassiterite has an extreme relief, stronger colours and birefringence. The RI of xenotime is lower and it exhibits higher-order, almost uniform white, polarization colours. Xenotime can also be distinguished by its absorption spectrum.

Radiation damage, resulting from radioactive thorium and uranium in zircon, converts crystalline zircon into an optically amorphous structure. The process is termed metamictization and **metamict** or **malacon zircons** are dusky dark grey, brown or black and usually isotropic.

Occurrence: Zircon is a remarkably widespread accessory mineral in rocks of crustal origin. It is particularly ubiquitous in silicic and intermediate igneous rocks. It also forms in mantle xenoliths, lunar rocks and meteorites. Zircon may reach high concentration in some beach sands and placers.

Remarks: No other accessory mineral has been studied as intensively as zircon. This can be attributed to special qualities such as its ubiquitous presence in almost all types of rocks, its morphological characteristics and high mechanical and chemical stability. Voluminous and exhaustive studies have focused on the above aspects (see Speer 1980 for reference).

Zircon morphology is widely used to determine the nature and genesis of the rocks. In intrusive granitoids zircon is generally euhedral. The character of relict zircons in metamorphic rocks (except migmatites) may be a key in indicating the igneous or sedimentary origin of the protolith. Rounded

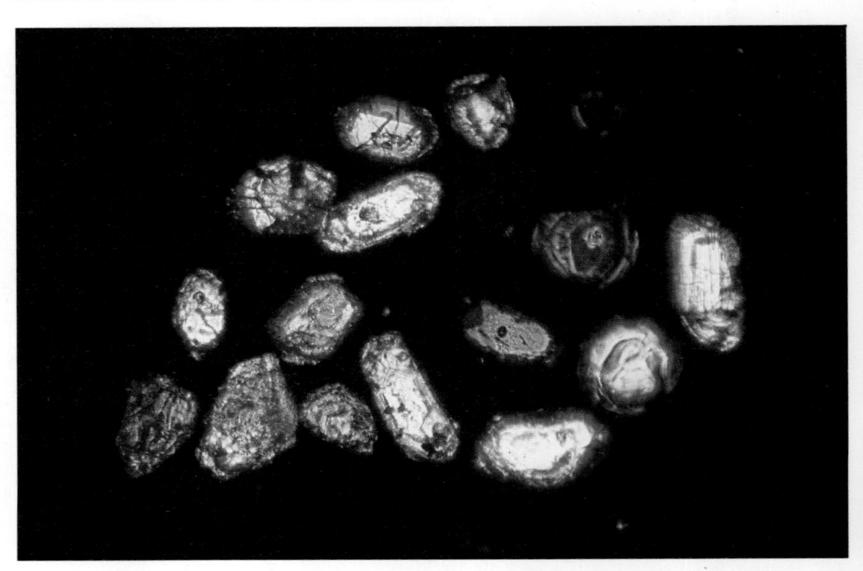

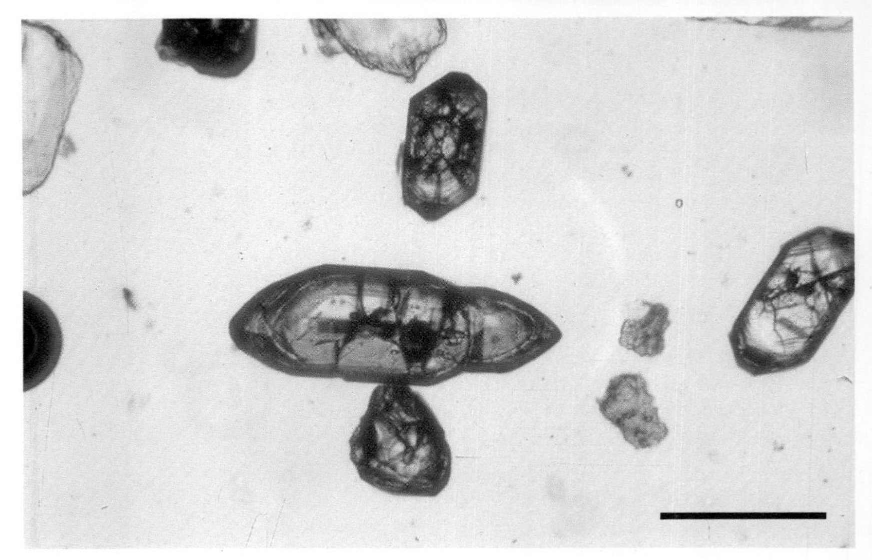

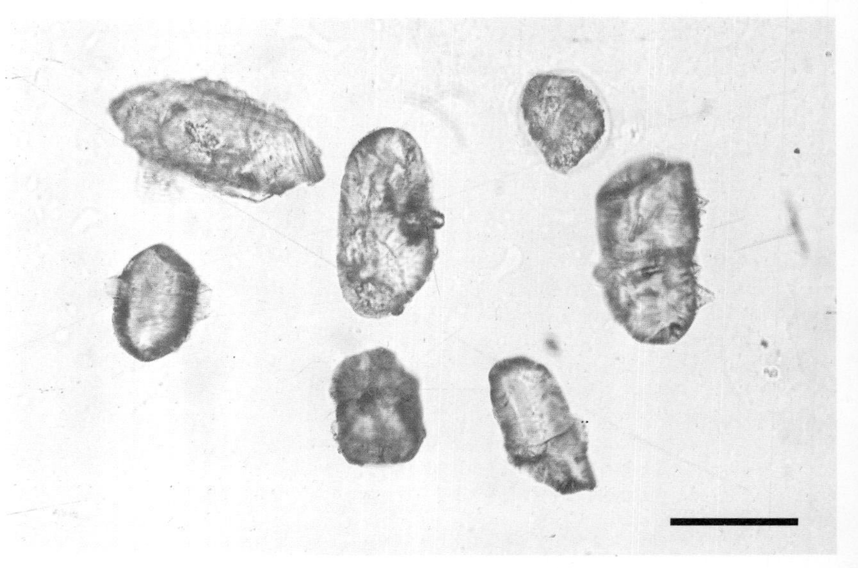

shape, especially high rounding index, is regarded as a criterion of detrital origin, though occasionally this may be caused by magmatic corrosion. The terminations of zircons of orthogneisses are often curved and zircons of kimberlites are rounded (Kresten *et al.* 1975). During intermediate and high-grade metamorphism, zircons retain their inherited shape; therefore rounded zircons may be supplied by various metasediments, schists and para-gneisses! If conditions are favourable during metamorphism, grains may develop overgrowth. In migmatites, anatexites and some contact-metamorphic rocks, the majority of zircons recrystallize to euhedral crystals but the presence of a relict core is not uncommon.

The nature of zircon in sedimentary and igneous rocks was highlighted by Poldervaart (1955, 1956). Extensive studies on zircon morphology ('typologie') and implications were made by Pupin (1976), and Pupin & Turco (1972, 1981; see also the many references in the latter).

The purple colour of zircon is attributed to prolonged radiation bombardment; consequently the intensity of shades increases with the amount of radioactive elements in a grain and also with geological age. Studies by Mackie (1923b), Tomita (1954) and Zimmerle (1972) assumed that most purple zircons of sediments were derived from Precambrian source rocks. Gastil *et al.* (1967) noted that hyacinth zircons lose their colour progressively with increasing metamorphism.

Zircon is regarded as *one of the most stable minerals*. The mineralogical maturity of heavy mineral assemblages is defined by the zircon–tourmaline–rutile (ZTR) index (Hubert 1962). Zircon persists through several sedimentary cycles and diagenesis, as well as through metamorphism. Its refractory nature is well known. The resistance of zircon is attributed to small size and lack of cleavage. Under certain conditions it is susceptible to alkaline leaching (Carroll 1953, Blumenthal 1958, Coleman & Erd 1961). Carroll listed several weaknesses which are due to zoning, fractures, inclusions and broken faces. Metamictization results in structural instability. Though these factors exist, they influence only a negligible proportion of zircon populations.

Zircon in palaeogeographic studies should be treated cautiously, as its shape and size may be uniform throughout a formation. Pronounced changes, however, may indicate different sources. Pupin (1976) demonstrated the value of zircons in tracing the source of detritus of the Permo-Triassic sandstones in south-east France. Zircon plays an important role in geochronology (Faure 1977).

Grains from: First page, upper: (a) Upper Eocene tuff horizon, Possagno, Italy; (b) Carboniferous, borehole Weiach 1294 m, Switzerland; (c) zircon with overgrowth, Carboniferous, borehole Weiach, 1431 m, Switzerland; Middle row: Carboniferous, borehole Weiach, 1431 m, Switzerland; Lower row: Buntsandstein, Triassic, borehole Kaisten, 106 m, Switzerland (Mmt 1.662).
First page, lower: Zircons exhibiting various degrees of rounding – selected from several ancient sedimentary environments (Mmt 1.662).
Second page, middle: Zircons exhibiting multiple stages of growths, Buntsandstein, Triassic, borehole Pfaffnau-1, 1822 m (Aroclor).
Second page, lower: Zircons with (presumably xenotime) outgrowths, Triassic, North Sea (Mmt 1.662).

Sphene (titanite)

$CaTi[SiO_4]$ (O, OH, F)
monoclinic, biaxial (+)

$n\alpha$	1.843–1.950
$n\beta$	1.870–2.034
$n\gamma$	1.943–2.110
δ	0.100–0.192
Δ	3.45–3.55

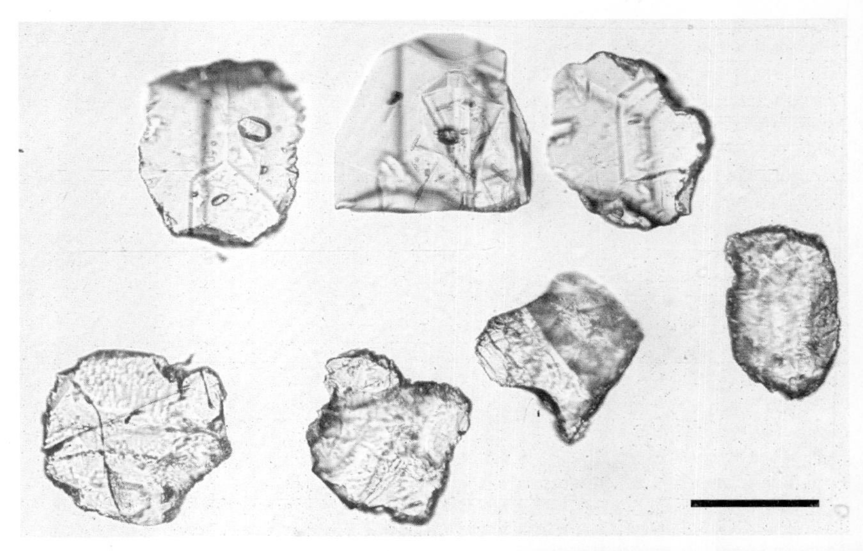

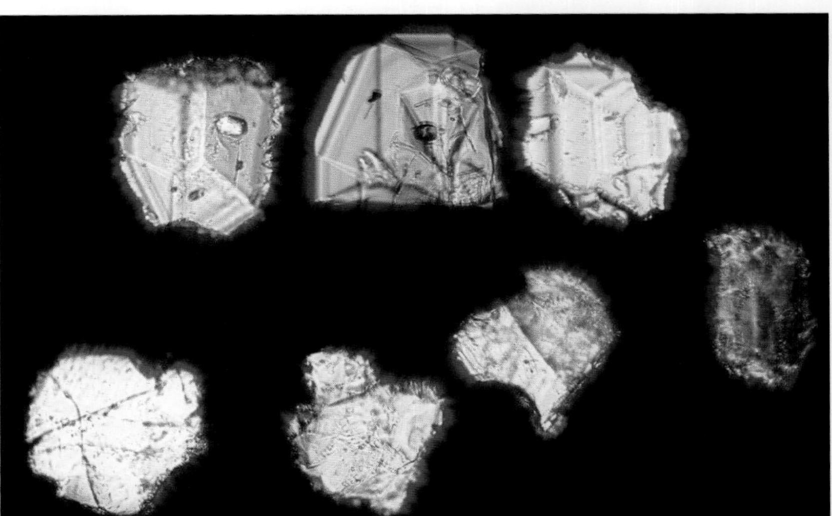

Form in sediments: Grains have a high resinous lustre and as a result of extreme RI their outline is surrounded by a dark rim. Sphene of pyroclastic origin is dominantly euhedral (upper row), but most detrital grains are irregular or rounded with a tendency to sphericity (lower row). Diamond-shaped grains may also be encountered. A certain degree of surface etching is common and etch-pits, facets and grooves, as well as fine hacksaw terminations, may be detected. Cleavages are seldom displayed. Grains often enclose fluid inclusions, carbonaceous matter, feldspar, quartz and zircon. Owing to alteration, sphene frequently appears dusky or may be partially converted to leucoxene (an aggregate of rutile, anatase and brookite) which is seen under reflected light as a yellowish-white substance.

Colour: Colourless, pale green, honey yellow or light brown.

Pleochroism: Coloured grains may show weak pleochroism: α colourless, pale yellow; β pale brownish yellow to pale yellowish green; γ orange-brown to pink or green.

Birefringence: Strong to extreme and the relief may show a slight 'twinkling' as the microscope stage is rotated. Interference colours are very high-order tints in golden yellow, yellowish white or pearl grey. Due to strong dispersion the colours often change to anomalous sky blue, which appears close to extinction point and is characteristic.

Extinction: Diamond-shaped grains have symmetrical extinction. Strong dispersion results in incomplete extinction.

Interference figure: Grains that show anomalous interference colours and incomplete extinction yield centred or slightly off-centre acute bisectrix figures and, depending on the Ti-content, somewhat variable 2V. The isogyres become broader or diffuse towards the exterior of the figure and are surrounded by numerous thin isochromatic curves. $r>v$ dispersion is extremely strong.

Elongation: Observation of the elongation is difficult as colours are obscured by high-order interference tints. The diamond-shaped grains are length fast in the direction of the long diagonal.

Distinguishing features: Sphene is diagnosed by high lustre, extreme relief, strong birefringence and usually abnormal polarization colours. The characteristic interference figure and strong dispersion is also diagnostic. Monazite, cassiterite, rutile, zoisite and staurolite may resemble sphene. However, monazite has lower RI and lower-order interference colours. The morphology of cassiterite is either prismatic or irregular; its colours are deeper and often appear in a patchy arrangement. Zoisite has a lower relief and it displays markedly deeper anomalous interference colours. Staurolite is distinctly pleochroic, has lower refractive indices, weaker birefringence and normal polarization tints. Rutile exhibits a higher relief, distinct pleochroism and normal interference colours.

Occurrence: Sphene is a widespread accessory mineral of undersaturated and intermediate plutonic rocks. It may form in pegmatites and in low-temperature Alpine-type veins, also in skarns. It is less common in volcanic rocks. Metamorphic schists, granite-gneisses, amphibolites and metamorphosed impure calc-silicate rocks commonly contain sphene.

Remarks: Sphene is chemically unstable and usually dissolves at an early stage of diagenesis. In special cases it decomposes into leucoxene, quartz and calcite (Morad & Aldahan 1985) or forms minute authigenic crystals together with authigenic titanium minerals (Gorbatschev 1962, Tröger 1969, Aldahan & Morad 1986).

Grains from: Upper row: Tertiary infilling of bauxite pockets, Seydisehir, Turkey; Lower row: Carboniferous, North Sea (Mmt 1.662).

Garnet group

isometric

Pyralspite garnets

		n	Δ
almandine	$Fe^{2+}Al_2[SiO_4]_3$	1.830	4.1–4.3
spessartine	$Mn_3Al_2[SiO_4]_3$	1.800	3.8–4.3
pyrope	$Mg_3Al_2[SiO_4]_3$	1.714	3.5–3.8

Ugrandite garnets

		n	Δ
grossular	$Ca_3Al_2[SiO_4]_3$	1.734	3.4–3.6
andradite	$Ca_3Fe_2^{3+}[SiO_4]_3$	1.887	3.7–4.1
uvarovite	$Ca_3Cr_2[SiO_4]_3$	1.860	3.4–3.8

Form in sediments: Detrital garnets occur as euhedral crystals (upper, first row), sharp irregular fragments (middle, upper row) and subrounded to rounded grains (middle, middle row). Uneven or conchoidal breakage patterns are often exhibited. Dissolution features such as surface pitting, mamillae and well-formed etch-facets frequently develop on grains during diagenesis (middle, lower row). Cleavage is absent but {100} parting on almandine may produce platy grains. Inclusions of quartz, iron ore, apatite, zircon, rutile, muscovite, biotite and graphite are common.

Colour: Pyralspite garnets are colourless, light pale pink, pale red or pinkish brown. Grossular of the ugrandite group is colourless to pale green, andradite is pale to deep brown and uvarovite is pale green.

Birefringence: Pyrope and almandine are truly isotropic, spessartine may show slight anisotropism, and the ugrandite group, especially grossular and andradite, commonly have weak birefringence and show grey or bluish-white interference colours (lower).

Interference figure: Anisotropic garnets are uniaxial or biaxial negative with variable 2V and sometimes strong dispersion. Contact metamorphic garnets may show alternating isotropic and birefringent zones parallel to the crystal faces. 'Pie-shaped' wedges or sector twinning (lower, right) may be visible on detrital grains.

Distinguishing features: Garnet is common in sediments and its high relief, together with isotropic character, enables an easy diagnosis. Etch facets, when present, are also distinctive. Birefringent garnets may cause confusion, but their low interference colours and zoning or twinning assist in the identification. The colour of spinels is deeper, sometimes they display octahedral parting and are truly isotropic.

Occurrence: Garnet is common in a variety of metamorphic rocks and is also present in plutonic igneous rocks, pegmatites, in ultramafic varieties and in some acid volcanics. Ca-garnets (ugrandites) are restricted to contact metamorphic impure limestones, metasomatic skarns and to schists formed from impure limestones by regional metamorphism. Uvarovite is a rare species and is found in peridotites or serpentinites. In sediments almandine is the most widespread garnet.

Remarks: The chemical composition of garnets is a function of the chemical and P–T conditions during crystal growth; therefore a knowledge of detrital garnet chemistry and

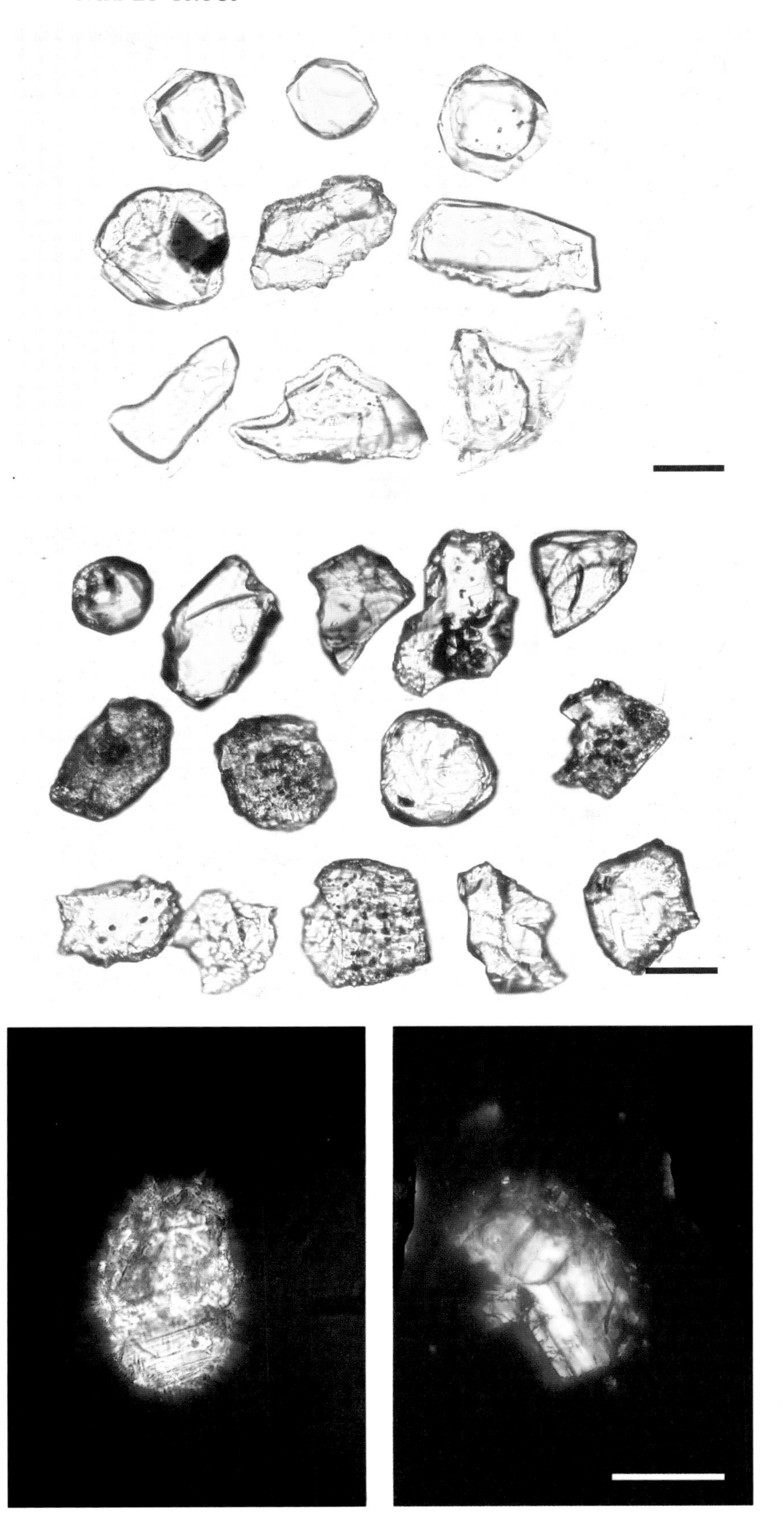

compositional zoning is a reliable tool in reconstructing source rock types and provenance (Morton 1985b). Garnets are fairly sensitive to acid leaching (see p. 7) but they can survive recycling and, to a certain degree, burial diagenesis. In deeply buried rocks, faceted garnets are often found together with highly resistant zircon, tourmaline, rutile and, not uncommonly, apatite.

Grains from: Upper: Upper row: stream sand, Melezza, Switzerland; Middle row, left: Oligocene glass sand, Laussach, southern Germany; middle: Oligocene, Barrême Basin, France; right: Carboniferous, North Sea; Lower row: Oligocene, Barrême Basin, France (Mmt 1.662).
Middle: Upper row: Carboniferous, North Sea; Middle row: Borehole Leymen-1, 252 m France; Lower row: etched grains, Devonian, North Sea (Mmt 1.582).
Lower: anisotropic and twinned grains, Oligocene Molasse, Savoy, France (Mmt 1.662).

Vesuvianite (idocrase)

$Ca_{19}(Al,Fe^{3+})_{10}(Mg,Fe^{2+})_3(Si_2O_7)_4(SiO_4)_{10}(O,OH,F)_{10}$
tetragonal, uniaxial (−)

$n\varepsilon$	1.700–1.746
$n\omega$	1.703–1.752
δ	0.001–0.008
Δ	3.33–3.43

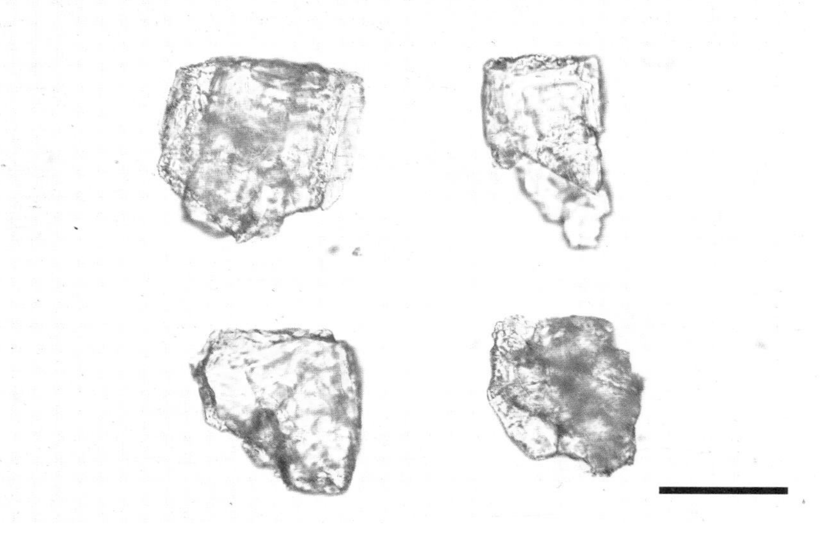

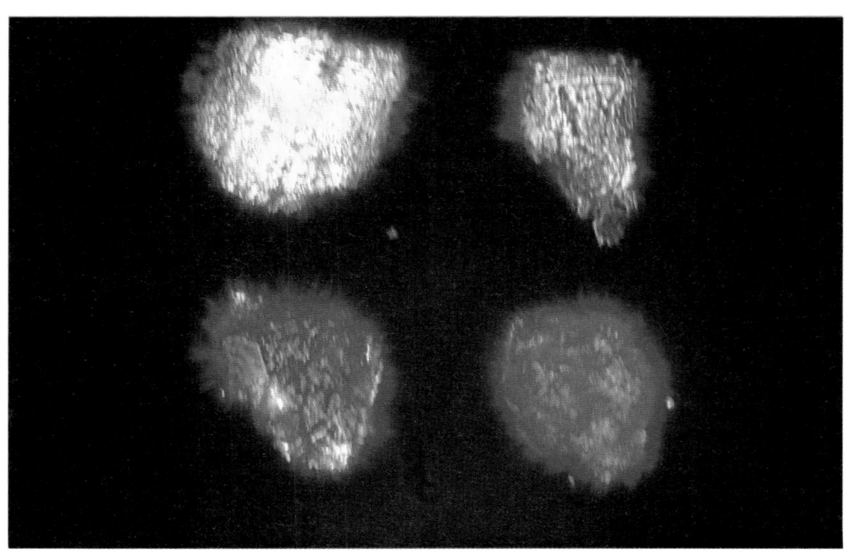

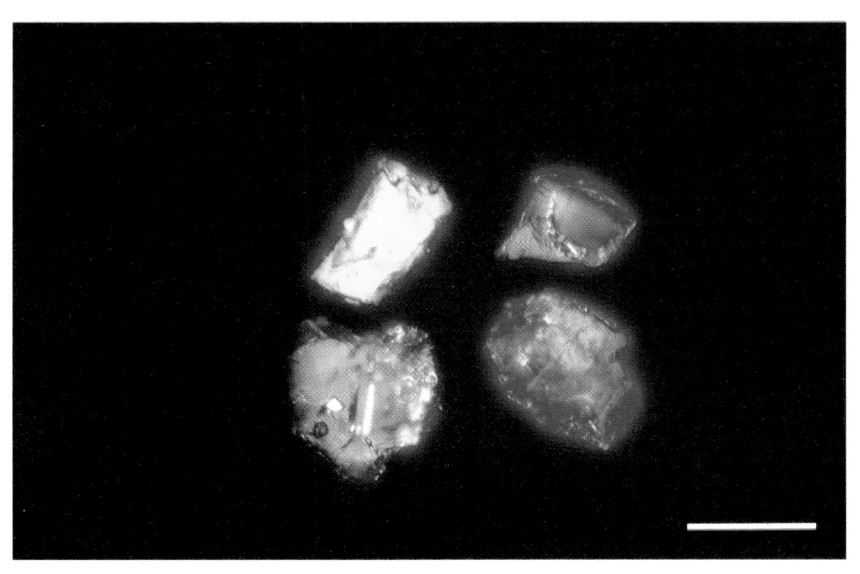

Form in sediments: Vesuvianite occurs as short or, more rarely, long prisms, prismatic fragments; also sharp angular, irregular or equant grains. Rounded corners and rounded to well-rounded forms are not uncommon. Cleavages are seldom observed. The prismatic fragments are usually bound with plain, even crystal faces and their surfaces may show small etch pits, cracks or dark discolouration. Inclusions are infrequent.

Colour: Generally colourless, pale yellow or pale green. Thicker grains occasionally display colour zoning or irregular colour distribution.

Pleochroism: Coloured grains may show faint pleochroism.

Birefringence: Weak, and grains exhibit first-order deep-yellow or dull greyish-yellow interference colours which are often mottled near extinction point. Most striking are, however, the anomalous interference tints, which range from deep dark yellow, brownish yellow through purple to intense blue (Berlin blue) or dark mauve. Some crystals show homogeneous polarization colours, others display a range of the above shades, often varying within a single grain. Zonal arrangement of the yellow or anomalous colours is occasionally detectable.

Extinction: Prismatic grains have parallel extinction.

Interference figure: Often poorly displayed. Many grains show faint flashes of biaxial isogyres with large 2V in a dull dark brownish-yellow field. Well-defined isogyres and distinct dispersion may be discernible on some specimens and more rarely an off-centre uniaxial figure is seen. The biaxial crystals are either optically positive or negative, but uniaxial grains generally show a negative sign.

Elongation: Prisms display negative elongation.

Distinguishing features: High relief and characteristic polarization colours are diagnostic. However, the latter may lead to confusion, mistaking vesuvianite for zoisite, clinozoisite or sphene. Clinozoisite has good cleavages, but these are rarely displayed by vesuvianite. The yellow shades of the zoisite–clinozoisite interference tints are of higher order and the deep purple–mauve and Berlin blue colours are absent. Sphene has a higher relief, a distinct surface reflection and strong dispersion. Birefringent grossular garnet, which is often associated with vesuvianite, has greyish-white interference colours and strongly mottled or undulatory extinction.

Occurrence: The principal parent rocks of vesuvianite are contact metamorphic parageneses associated with impure limestones. Vesuvianite also occurs in skarns and occasionally in greenschists. As a result of calcium metasomatism it forms veins in ultrabasic rocks, in serpentinites and it is a characteristic mineral of rodingites.

Remarks: Vesuvianite in sediments has been reported from various localities (see Milner 1962). It was also mentioned by Gazzi & Zuffa (1970) from the Palaeogene Emilia sandstones of the northern Apennines and it is common in some Oligocene formations of the Barrême Basin, southern France (Evans & Mange-Rajetzky 1991).

Grains from: Upper: Oligocene, Barrême Basin, France (Mmt 1.582); Lower: same locality (Mmt 1.662).

Sillimanite

$Al_2O[SiO_4]$
orthorhombic, biaxial (+)

$n\alpha$	1.654–1.661
$n\beta$	1.658–1.662
$n\gamma$	1.673–1.683
δ	0.020–0.022
Δ	3.23–3.27

Form in sediments: Sillimanite most commonly occurs as long slender prisms elongated on *c*, short stout prismatic fragments, equant grains and also as cleavage fragments or irregularly shaped particles of larger crystals. Grains may display a certain degree of rounding. Beside the prismatic morphology, sillimanite crystallizes in a fibrous style which is usually called **fibrolite**. The latter appears as fine needles in parallel or subparallel intergrowth and also as interlacing mats of fibres (F). Occasionally both habits can be observed in a single grain (SF). Good {010} cleavage is exhibited by many specimens. Inclusions of spinel, biotite and zircon are often present.

Colour: Prismatic grains are colourless. Fibrolite appears with a pale green or pale brown hue.

Pleochroism: Slight pale brown to pale yellow pleochroism is rarely detectable on fibrous grains.

Birefringence: Moderate to fairly strong. A slight 'twinkling', especially of the fibres, is noticeable on rotation. Prisms and fibres display brilliant second- and third-order interference colours with yellow, green and deep pink as dominant shades. Basal fragments have low birefringence, hence their polarization colours are first-order grey and yellow.

Extinction: Of the prisms and fibres is parallel.

Interference figure: Prismatic grains and cleavage fragments yield flash figures. A well-centred acute bisectrix figure can be obtained from fragments and short prisms which lay on their basal plane or rolled onto their basal plane in order to obtain an interference figure. Isogyres appear in a white or yellow field without isochromes. 2V is moderate.

Elongation: Positive.

Distinguishing features: Prismatic habit and brilliant interference colours are fairly distinctive. Enstatite prisms with similar relief may be confused with sillimanite, but the former are usually broader, tabular and show lower birefringence. In addition they often exhibit exsolution lamellae, serpentinization and commonly appear with ragged edges or display other signs of dissolution. Thin sillimanite prisms with weaker polarization colours resemble apatite, but the latter is length fast. Andalusite has a lower-relief, an irregular habit, it is length fast and usually pleochroic. Anthophyllite prisms may be confused with sillimanite, but the former may be optically negative. Fractured or rounded grains are more difficult to identify. Fibrolite strongly resembles the fibrous anthophyllite variety gedrite, from which distinction by optical means alone is not always successful. Gedrite has deeper pale brown or clove brown colours, and usually displays weak to strong pleochroism. Staurolite, kyanite and andalusite in a heavy mineral suite may indicate the presence of sillimanite.

Occurrence: Sillimanite crystallizes in high-temperature metamorphic rocks and occurs in sillimanite–cordierite gneisses and biotite–sillimanite hornfelses. It is also present in granulite facies rocks. High-grade regional metamorphism of pelitic rocks also produces sillimanite.

Grains from: Grab sample, Port of Beira, Mozambique (Mmt 1.582).

Andalusite

$Al_2O[SiO_4]$
orthorhombic, biaxial (−)

$n\alpha$	1.629–1.640
$n\beta$	1.633–1.644
$n\gamma$	1.638–1.650
δ	0.009–0.010
Δ	3.13–3.16

Form in sediments: The morphology of detrital andalusite is angular irregular, occasionally prismatic, subrounded and rarely well rounded. Cleavage traces are poorly displayed. Grains frequently enclose carbonaceous impurities. Andalusite is fairly large and is usually found in the coarser grades.

Colour: Colourless or it has a pinkish tinge. The rare manganoan variety **viridine** is green and pleochroic in shades of emerald green and yellowish green.

Pleochroism: Some grains are non-pleochroic (lower row, left), others display a distinctive pleochroism: α, rose pink or yellow; β and γ, colourless, pale yellow, greenish yellow. The pattern of pleochroism is usually inhomogenous and there are non-pleochroic and strongly pleochroic patches within a single grain.

Birefringence: Weak. Interference colours are first-order grey and white, but these are seen only on thin fragments. The majority of andalusite grains display second-order orange, red and greenish-blue interference colours. Owing to the irregular shape and uneven thickness of most grains, the interference colours show a disorderly pattern. On rounded grains the concentric interference bands are similar to those of detrital quartz.

Extinction: Parallel to cleavage traces and crystal faces, but difficult to observe because of the irregular shape of the grains.

Interference figure: Basal fragments yield an acute bisectrix figure with large 2V and weak dispersion. Cleavage fragments show a flash figure.

Elongation: Negative.

Distinguishing features: Pleochroic andalusite can be easily recognized by its distinctive pleochroism, coupled with a moderate relief. Piemontite has high RI and is biaxial positive. Hypersthene is mostly prismatic, it exhibits a higher relief, pyroxene cleavages and positive elongation. Sillimanite commonly occurs as well-defined prisms or fibrous aggregates; it is non-pleochroic, biaxial positive and length slow. Andalusite, which is free of carbonaceous inclusions and lacks pleochroism, can be confused with topaz. However, the latter generally yields a good interference figure with a positive sign.

Occurrence: Andalusite is a typically metamorphic mineral and appears most commonly in argillaceous rocks of contact aureoles around igneous intrusions. It is common in gneisses and schists.

Remarks: **Chiastolite** is an andalusite variety, generated at relatively low temperatures. It contains abundant graphite and

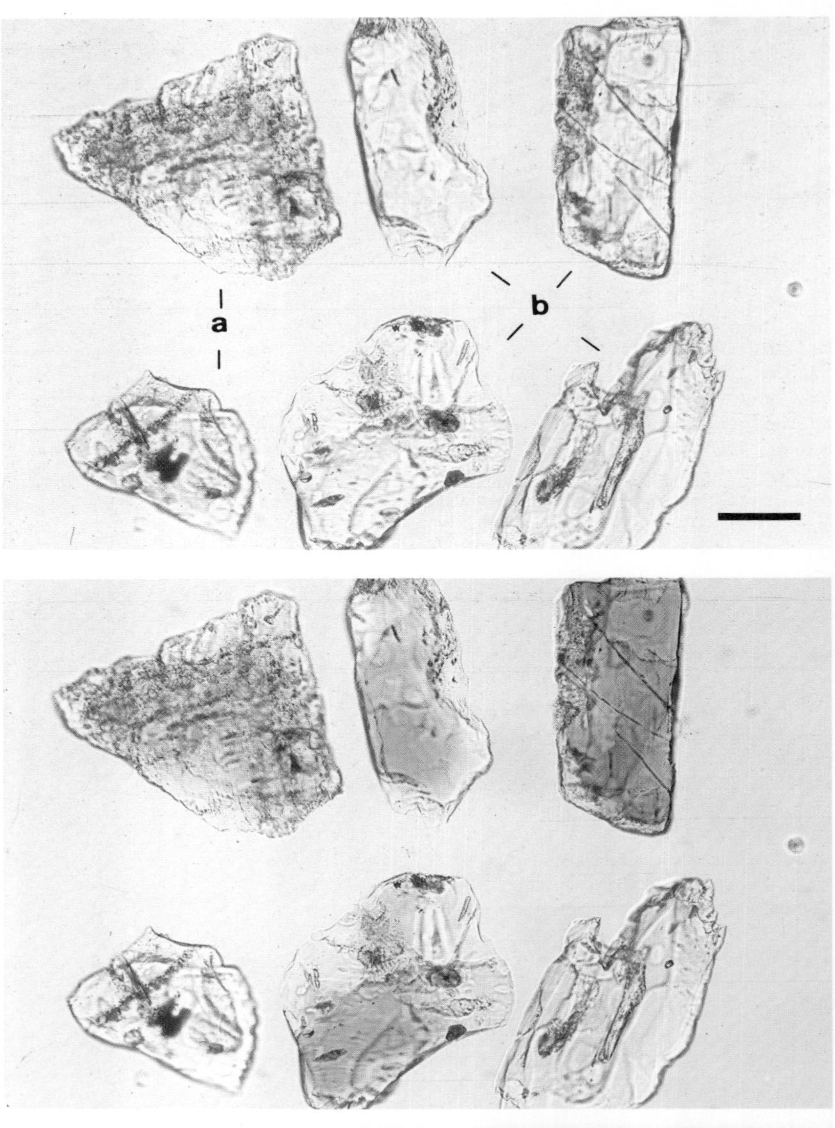

carbonaceous inclusions which are arranged systematically within the mineral, appearing as dark bands or as a black cross in a basal section. The form of detrital chiastolite is irregular or rounded, sometimes prismatic. Rarely basal fragments occur. It is colourless and non-pleochroic.

Grains from: (a) Aquitanian, Nerthe-Chain, Provence, France; (b) glass sand from the Oligocene Molasse, southern Germany (Mmt 1.582).

Kyanite (disthene)

$Al_2O[SiO_4]$
triclinic, biaxial (−1)

$n\alpha$	1.712–1.718
$n\beta$	1.721–1.723
$n\gamma$	1.727–1.734
δ	0.012–0.016
Δ	3.53–3.65

Form in sediments: Kyanite grains may be quite large and are more frequent in the coarser grades, but small cleavage flakes also appear in fine-grained sediments. Grains are most commonly angular, bladed or prismatic and are elongated on *c*. They usually lie on the {100} face, which is the plane of perfect cleavage. Resulting from the combination of {100} cleavage and parting on {001} at almost right angles to the length of the crystals, grains exhibit characteristic cross fractures and step-like features. Rounding of the corners may be frequent, or occasionally well-rounded forms appear. There are square-shaped fragments showing no cleavage traces and also some which are bent or irregularly shaped. Intrastratal solution produces fine etch-facets on the edges of the grains and on the parting planes. Fluid inclusions, also rutile, apatite, graphite, etc., are common.

Colour: Grains are dominantly colourless or rarely blue, often with an uneven colour distribution.

Pleochroism: Blue-coloured specimens have weak pleochroism: α, colourless; β, violet blue; γ, cobalt blue.

Birefringence: Moderate and thin cleavage flakes exhibit first-order grey and yellow interference colours. These may range to second-order orange, purple and blue in thicker grains. The step-like uneven thickness often results in a spectacular arrangement of the interference colours, appearing pale grey on the thinner edges and deep purple or blue in the thicker parts.

Extinction: The extinction angle measured on {100} is large, ranging between 27°–32°. Grains laying on {010} have parallel extinction.

Interference figure: {100} sections provide well-centred or nearly centred acute bisectrix figures with large 2V. Isogyres appear on a white or yellow field with or without isochromes. Dispersion is weak.

Elongation: Positive.

Distinguishing features: Habit, the combination of perfect cleavage and parting, together with a large extinction angle and a good interference figure, provide an easy diagnosis. The identification of rounded and highly etched irregular grains is more difficult. In these cases one of the diagnostic characteristics may be visible, leading to recognition. The large extinction angle assists in distinguishing kyanite from sillimanite. Andalusite has a lower relief and it is length fast. Some baryte cleavage fragments resemble kyanite but have lower RI and smaller extinction angle.

Occurrence: Kyanite occurs in gneisses, granulites and pelitic schists which are generated by the regional metamorphism of mostly pelitic rocks. It is considered as an indicator of the metamorphic zone higher than the one in which staurolite forms but preceding sillimanite grade.

Remarks: Kyanite is a common detrital mineral. It is resistant to acidic weathering (Grimm 1973, Nickel 1973) and is fairly stable in the diagenetic environment. With increasing depth of burial it dissolves before staurolite (Morton 1984b).

Grains from: (a) Grab sample, Port of Beira, Mozambique; (b) Lower Miocene Molasse, Valensole Basin, France (Mmt 1.662).

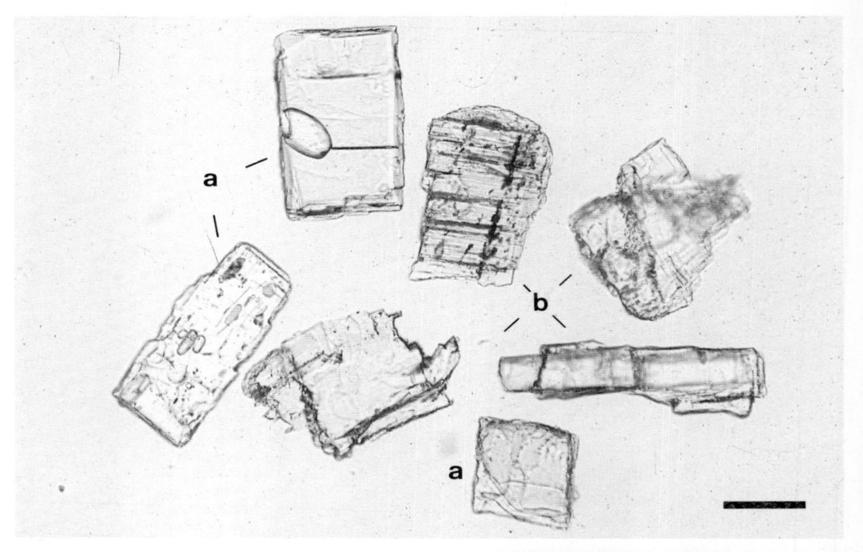

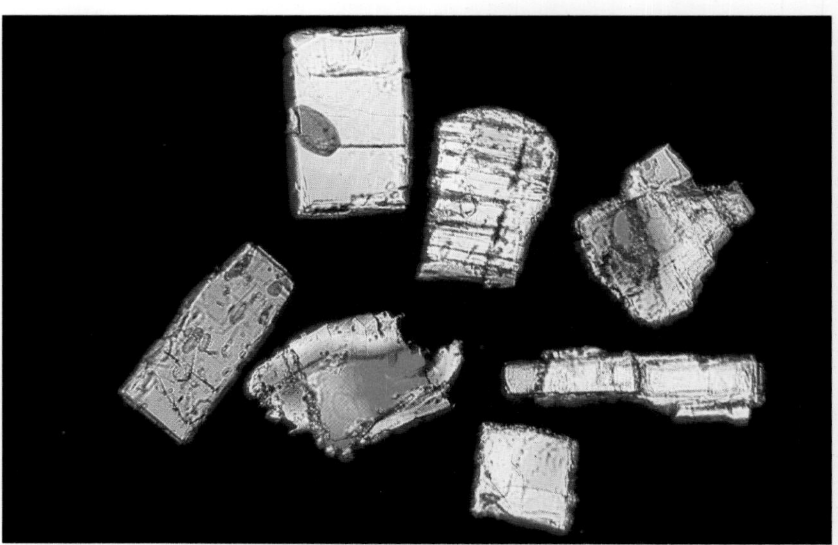

Topaz

$Al_2O[SiO_4](F,OH)_2$
orthorhombic, biaxial (+)

$n\alpha$	1.606–1.629
$n\beta$	1.609–1.632
$n\gamma$	1.616–1.638
δ	0.008–0.011
Δ	3.49–3.57

Form in sediments: Topaz appears as angular irregular or sometimes rounded grains. Basal fragments usually display traces of perfect {001} cleavage. The surface of the grains is commonly marked by crescent-shaped indentations. Fluid or opaque inclusions are frequent.

Colour: Most detrital grains are colourless with a noticeable bluish-white tinge, but pale-coloured yellow, pink, red and blue grains are also reported.

Pleochroism: Absent or very weak on thick grains: α, pale yellow or brown; β, pale orange; γ, pale pink.

Birefringence: Weak, and thin fragments yield first-order grey, white or pale yellow interference colours. Thick detrital grains show a range of colours up to second-order orange, red, blue and green.

Extinction: Parallel to cleavage traces and to prismatic faces and symmetrical on basal sections.

Interference figure: Grains laying on their basal plane yield a centred acute bisectrix figure with a distinct r>v dispersion. Many detrital topaz can be diagnosed with certainty by rolling them in a liquid immersion medium onto their basal plane, which then will provide a clear interference figure. Grains orientated according to the basal plane display lower-order interference colours (lower, (a) right-hand grain).

Elongation: Positive to crystal elongation, and negative to cleavage traces.

Distinguishing features: Bluish-white tinge and an irregular or subrounded form with frequent indentations and perfect basal cleavage are characteristic. It can be easily mistaken for non-pleochroic andalusite but the latter is optically negative (though it often fails to show a discernible interference figure). Etched, rounded forsteritic olivine grains resemble topaz; however their RI and birefringence are considerably higher. The relief of apatite is higher, it is uniaxial and length fast. Quartz has lower refractive indices. Untwinned lawsonite, which also exhibits a bluish tinge, similar interference colours, extinction and inclusions, can be confused with topaz, but lawsonite grains are thinner, platier, have a higher relief and larger 2V. It may be difficult to distinguish sericitized topaz from sericitized andalusite.

Occurrence: Topaz is formed primarily in granite, granite pegmatite and in greisen. High-temperature veins and ore deposits as well as vugs, cavities and fissures of acid rocks may also contain topaz. It occurs in metamorphic rocks formed by F-metasomatism and rarely in muscovite–kyanite-schists.

Grains from: (a) Lower Miocene Molasse, Anzfluh, Switzerland; (b) beach sand, St Ives, England (Mmt 1.539).

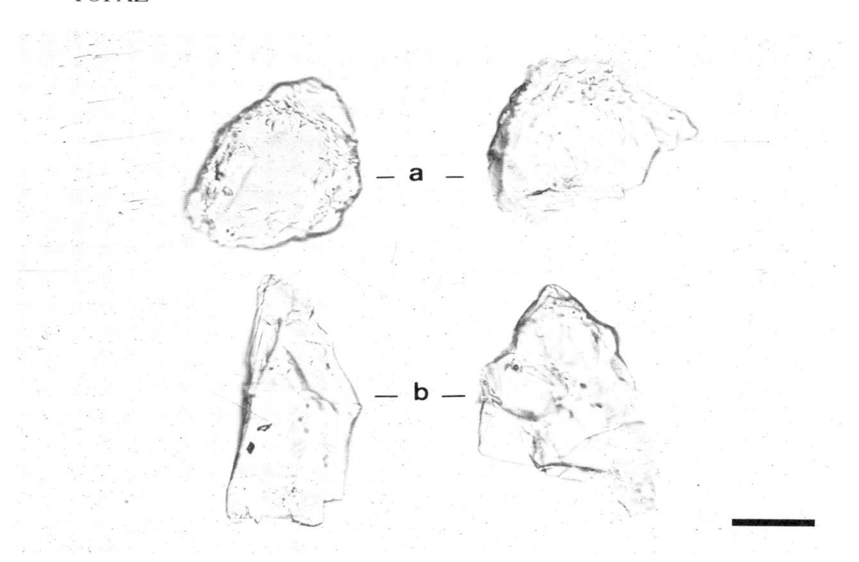

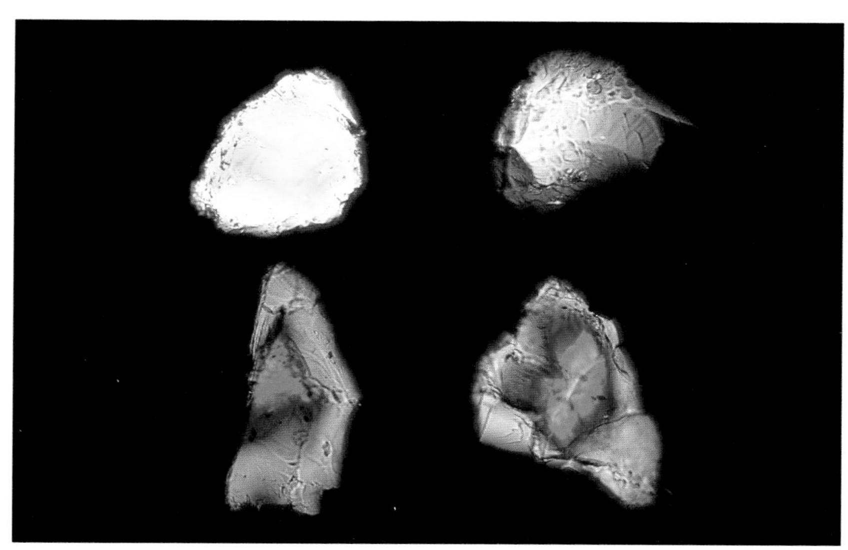

Dumortierite

$(Al,Fe^{3+})_7O_3(BO_3)[SiO_4]_3$
orthorhombic, biaxial (−)

$n\alpha$	1.655–1.686
$n\beta$	1.675–1.722
$n\gamma$	1.684–1.723
δ	0.010–0.037
Δ	3.30

Form in sediments: Detrital dumortierite is generally short prismatic or elongated according to its long axis. Less commonly, irregular fragments appear. Striations parallel with the length of the prisms may be exhibited by some specimens. The edges of the grains occasionally show rounding. Distinct {100} cleavages are occasionally visible.

Colour: Blue, greenish blue, lavender, pink, reddish violet.

Pleochroism: Strong and distinctive. It appears in contrasting shades of deep blue, indigo or red-violet: α, blue, violet, greenish blue; β, colourless, red-violet, pale lilac, yellow; γ, colourless, pale blue, pale green. Maximum absorption appears when the length of the crystals is parallel to the vibration direction of the polarizer.

Birefringence: Weak to moderate, and maximum interference colours range from first-order white, orange and pink to second-order blue and yellow on thick fragments.

Extinction: Parallel to the length of the prisms or striations.

Interference figure: {100} cleavage fragments yield obtuse bisectrix figures with poorly defined isochromatic curves.

Elongation: Negative.

Distinguishing features: Moderate relief, colour, very intensive pleochroism and negative elongation are diagnostic. Strongly pleochroic tourmaline is liable to be confused with dumortierite, but tourmaline is uniaxial and its maximum absorption is normal to the vibration direction of the polarizer. This is a reliable property that assists in distinguishing between these two minerals. Sodic amphiboles have inclined extinction and amphibole cleavages. The RI of corundum is higher and it is uniaxial. Piemontite has a distinctly stronger relief and a positive sign.

Occurrence: Dumortierite occurs in granite pegmatites, aplites and quartz veins. It is also present in some quartzites, in granite gneisses and micaschists, as well as in hydrothermally altered quartz-feldspar rocks.

Grains from: Upper: Lower Greensand, Cretaceous, Dorking– Leith Hill District, England (CB). From the collection of H. B. Milner.
Middle: Wealden Basin, Bristol Channel, England (Mmt 1.582). Courtesy of A. C. Morton.

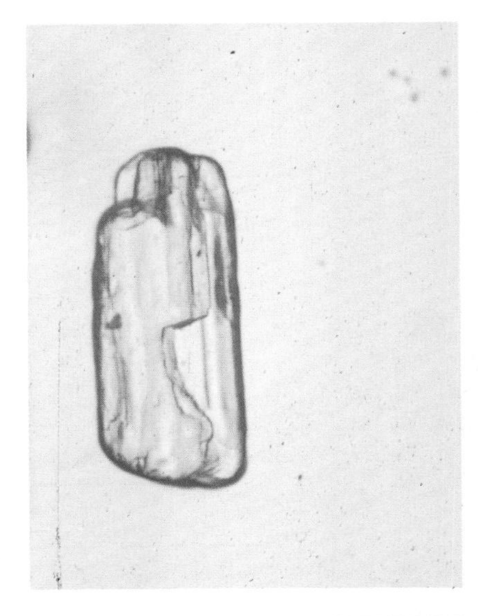

Staurolite

$(Fe^{2+}Mg)_2(AlFe^{3+})_9O_6[SiO_4](O,OH)_2$
monoclinic, pseudo-orthorhombic, biaxial (+)

$n\alpha$	1.739–1.747
$n\beta$	1.745–1.753
$n\gamma$	1.752–1.761
δ	0.012–0.014
Δ	3.74–3.83

Form in sediments: Staurolite occurs as irregular, angular, somewhat platy, often fractured grains which show poorly defined cleavage traces. Rarely, diamond-shaped basal sections are found. Ragged outline, etch patterns and mamillae are frequently exhibited by grains which have experienced dissolution during diagenesis (d). Inclusions of quartz and carbonaceous matter are common (a).

Colour: Staurolite has bright yellowish colours in shades of pale yellow through golden yellow to dark yellowish brown.

Pleochroism: The intensity of pleochroism is influenced by the thickness of the grains but even thin pale-coloured fragments display noticeable pleochroism: α, colourless to pale yellow; β, pale yellow; γ, golden yellow to yellowish brown.

Birefringence: Moderate, and interference colours range from first-order grey or yellow in thin fragments to second-order orange, red and bluish green in thicker specimens.

Extinction: Owing to the irregular shape of the grains it is difficult to observe the extinction angle. Prismatic fragments and euhedral crystals show parallel extinction.

Interference figure: Basal sections (which can be recognized by usually lower-order polarization colours) yield an off-centre acute bisectrix figure on a white or pale yellow field with discernible r>v dispersion. Isochromes are generally absent. Other forms give poor eccentric figures.

Elongation: Prismatic grains are length slow.

Distinguishing features: Staurolite is one of the easily identifiable detrital minerals. High relief, combined with shades of yellow or yellowish brown, and distinct pleochroism are diagnostic. It may be confused with yellow or yellowish-brown tourmaline (T) but the latter has lower relief, is uniaxial negative and length fast.

Occurrence: Staurolite is almost exclusively a product of medium-grade regional metamorphism and it forms in mica schists, derived from argillaceous sediments, and less frequently in gneisses.

Remarks: Staurolite is a common detrital mineral. It is fairly resistant to weathering and burial diagenesis and, to a certain extent, recycling. In the sediments it is associated with either kyanite, andalusite and/or sillimanite, which points to their common parentage, or it is found in an impoverished mineral suite together with apatite and garnet as well as ultrastable zircon, tourmaline and rutile. The latter case may be the result of either recycling of mature pre-existing sediments (and this is indicated by the well-rounded associated minerals) or disso-

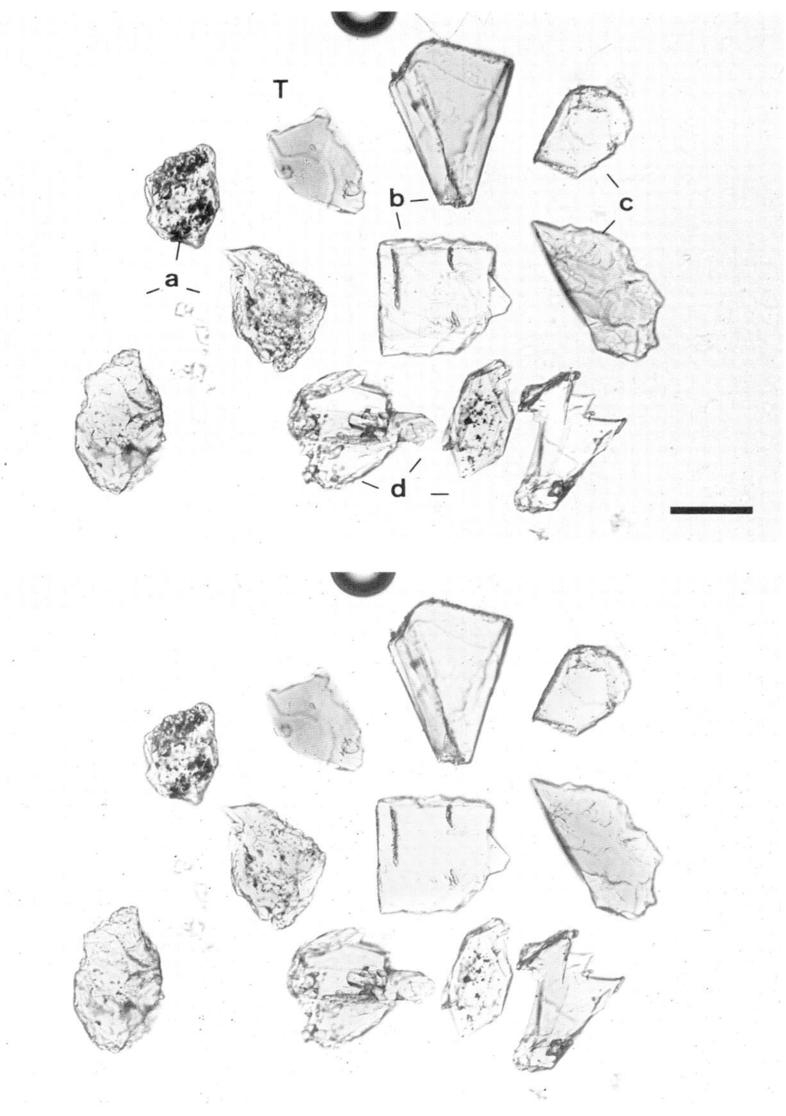

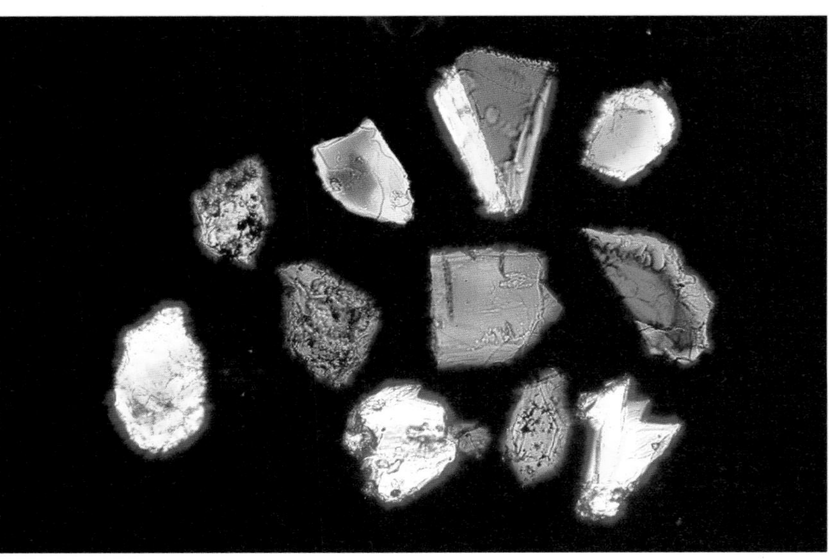

lution during diagenesis, eliminating the less stable species. The diminution and then the absence of staurolite with increasing depth as a function of intrastratal dissolution was shown by Grimm (1973) and Morton (1979a, 1984b).

Grains from: (a) Glass sand from the Oligocene Molasse, southern Germany; (b) grab sample, Port of Beira, Mozambique; (c) Tertiary, North Sea; (d) Carboniferous, North Sea; (T) tourmaline (Mmt 1.662).

Chloritoid

$(Fe^{2+},Mg,Mn)_2(Al,Fe^{3+})Al_3O_2[SiO_4]_2(OH)_4$
monoclinic, triclinic, biaxial (+) or (−)

$n\alpha$	1.713–1.730
$n\beta$	1.719–1.734
$n\gamma$	1.723–1.740
δ	0.006–0.022
Δ	3.51–3.80

Form in sediments: The layer-like nature of chloritoid results in a perfect mica-type {001} cleavage which determines its detrital appearance. Hence detrital grains occur dominantly as basal cleavage plates. A grain usually consists of several superimposed basal plates and shows a foliated habit. There is a tendency for the basal plates to curve. Chloritoid contains a wide variety of inclusions, mostly quartz, muscovite, rutile, tourmaline and garnet. Some grains are speckled with minute impurities. Although twinning is very common, it is rarely noticeable on detrital grains.

Colour: Chloritoid has a distinct greyish blue or slightly greenish to yellowish blue colour. Colour zoning is common.

Pleochroism: Distinct pleochroism is displayed in shades of grey and blue: α, pale grey, greyish green; β, slaty blue, indigo blue; γ, pale yellow, pale green.

Birefringence: Maximum birefringence is weak to moderate on {010} or {110} faces, but is very weak on the basal plane. Interference colours observed on the latter are first-order grey and white, sometimes anomalous, caused by inherent mineral colour. Thicker grains, lying on non-basal faces exhibit extreme pleochroism and second-order colours which may be confusing. The identity of such grains can be confirmed easily by turning them onto their basal plane in a liquid immersion medium.

Extinction: Varies from almost parallel to 18°. On the basal plane it is symmetrical or inclined to the prismatic {110} cleavages.

Interference figure: Basal plates yield a nearly centred acute bisectrix figure with varying 2V and strong r>v dispersion. The optical sign is usually positive.

Elongation: Negative.

Distinguishing features: Platy habit, colour, pleochroic scheme, relatively high relief combined with low birefringence on the basal plane, as well as the tendency to yield good interference figures – are diagnostic. These properties assist in distinguishing chloritoid from other foliated minerals. Chlorite has lower RI and small 2V, and its colours are in shades of green and bluish green. Prismatic fragments may resemble blue–green hornblende or tourmaline, but the former is optically negative and length slow, whereas tourmaline has a lower relief and is uniaxial.

Occurrence: Chloritoid is a product of low- to lower-medium-grade metamorphism and is generated by the regional metamorphism of pelitic sediments. It is regarded as a metamorphic stress mineral. In a non-stress environment it is formed by hydrothermal processes in veins and cavity fillings.

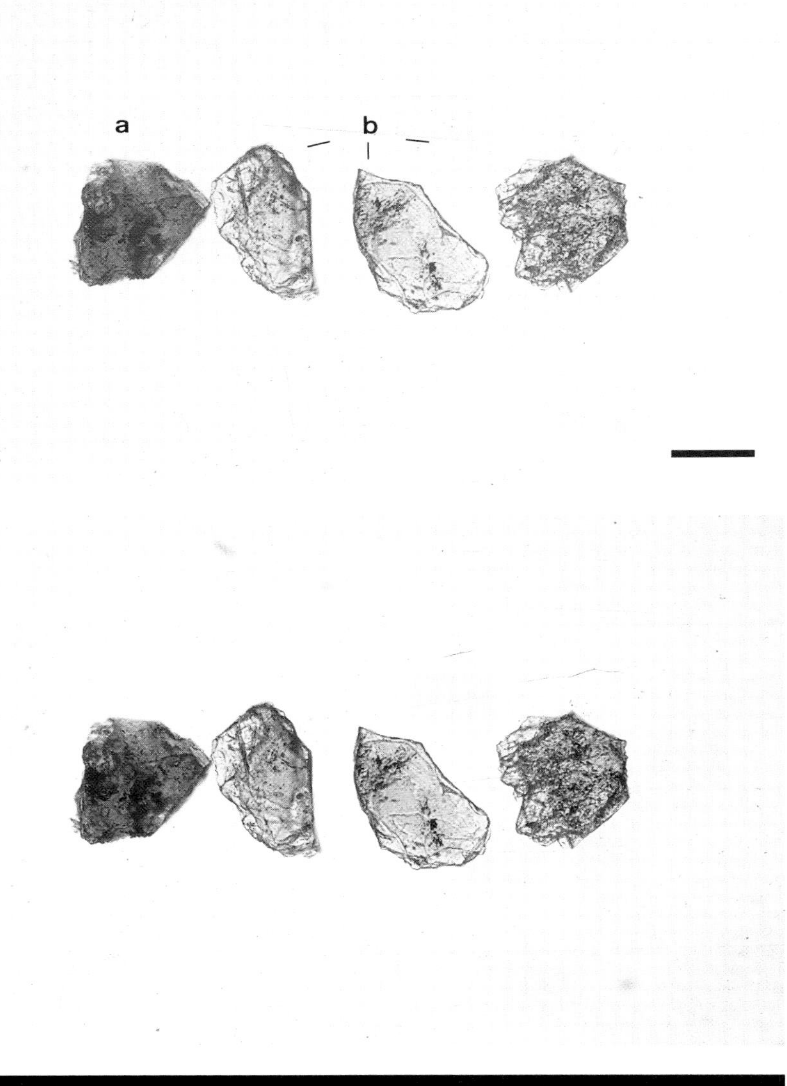

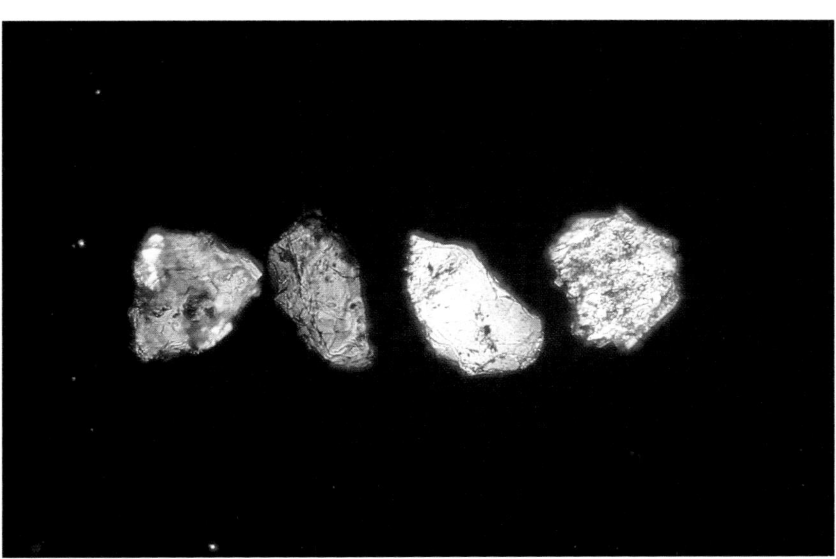

Grains from: (a) River sand, Danube, Hungary; (b) Lower Miocene Molasse, Anzfluh, Switzerland (Mmt 1.582).

EPIDOTE GROUP

Zoisite

$Ca_2Al.Al_2O.H[Si_2O_7][SiO_4]$
orthorhombic, biaxial (+)

$n\alpha$	1.685–1.705
$n\beta$	1.688–1.710
$n\gamma$	1.697–1.725
δ	0.004–0.008
Δ	3.15–3.27

Two varieties of zoisite have been distinguished: iron-free or α zoisite and ferroan or β zoisite. In accord with chemistry, their optical properties vary.

Form in sediments: Grains are short stumpy prisms or thin rectangular cleavage fragments, sometimes with a slightly rounded outline. {010} cleavage traces are often detectable. Compositional zoning is common. Opaque, fluid or needle-like amphibole inclusions may be present in some grains.

Colour: Common zoisite is colourless, whereas the rare manganoan variety **thulite** is pink.

Birefringence: Moderate. Changes of Fe^{3+} content may result in variations of birefringence within a grain. Iron-free zoisites display either greyish-white or, more frequently, anomalous deep indigo-blue and yellow interference colours due to strong dispersion. However, these are not seen in all crystal orientations. When rolling the grains in a liquid immersion medium under crossed polars, the anomalous colours usually appear on the basal plane. Ferroan (β) zoisites and thulite have stronger birefringence and normal second-order interference colours.

Extinction: Parallel to prism-outlines, but it may be undulatory or incomplete.

Interference figure: α zoisites, showing abnormal interference colours, usually yield centred or nearly centred acute bisectrix figures with small 2V and strong $r>v$ or $r<v$ dispersion. Well-defined isogyres separate the field into blue and yellow sectors and the absence of isochromatic curves is characteristic. β zoisite provides off-centre or optic-axis figures.

Elongation: α zoisite has either positive or negative elongation; that of β zoisite is positive.

Distinguishing features: The identification of α zoisite is relatively easy, as high relief and short prismatic habit, together with contrasting anomalous interference colours, are diagnostic. Clinozoisite grains that show parallel extinction and abnormal interference colours may be confused with zoisite, but the latter provides a characteristic interference figure and its interference colours are markedly deeper. Apatite has a lower relief and duller, greyish interference colours. The recognition of β zoisite is more difficult and it is likely to be mistaken for clinozoisite, or for other minerals with similar optical properties. Sillimanite has lower birefringence and interference colours in shades of orange, red and green, usually arranged in a concentric fashion which is not typical of zoisite. The RI of topaz is lower and its shape is dominantly irregular. Vesuvianite displays intense 'Berlin blue', purple or dull yellow polarization colours. Corundum has a higher relief and is uniaxial. Thulite resembles hypersthene, andalusite and piemontite, but it is very rare in sediments.

Occurrence: Zoisite is generated by medium-grade regional metamorphism and is a characteristic constituent of calc-silicate granulites and calcite-zoisite micaschists. It also occurs in regionally metamorphosed basic igneous rocks. Zoisite is a component of **saussurite**, formed by the alteration of calcic plagioclase. As a detrital species it is less common than epidote or clinozoisite.

Grains from: Upper Miocene, western Hungary (Mmt 1.662).

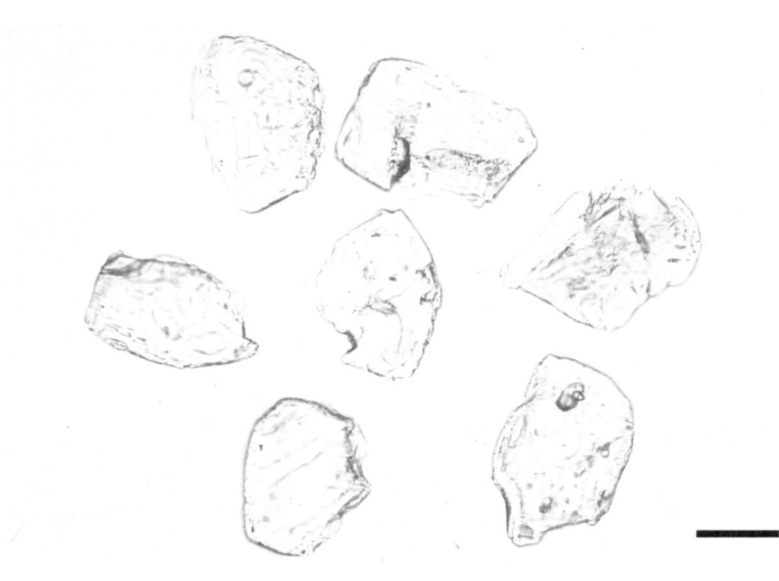

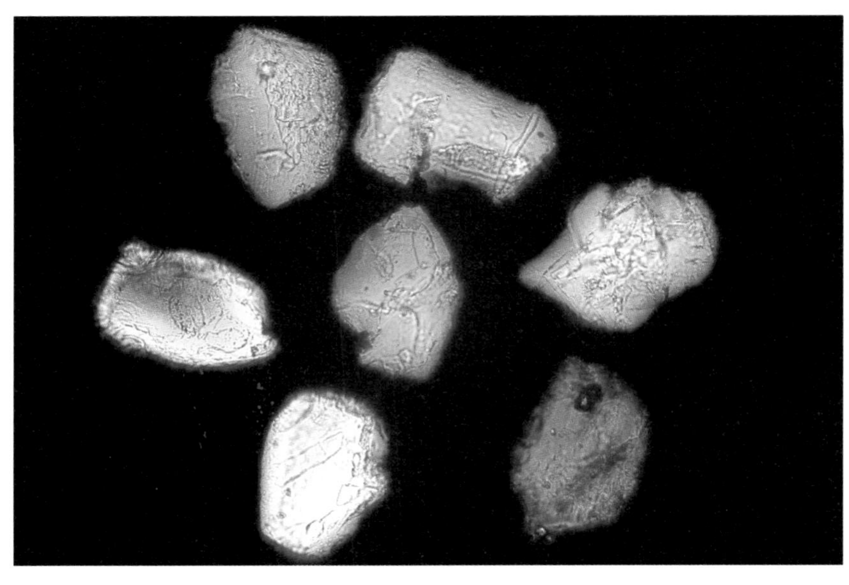

Clinozoisite

$Ca_2Al.Al_2O.OH[Si_2O_7][SiO_4]$
monoclinic, biaxial (+)

$n\alpha$ 1.670–1.715
$n\beta$ 1.674–1.725
$n\gamma$ 1.690–1.734
δ 0.005–0.015
Δ 3.12–3.38

Form in sediments: Clinozoisite has a high relief and it occurs as short or long prisms, broken euhedral crystals and rectangular fragments. Irregular or rounded grains, columnar and fibrous forms are infrequent. Perfect basal cleavage and zoning are occasionally discernible. Grains often enclose fluid or opaque inclusions. Etch patterns commonly develop on clinozoisite grains during diagenesis.

Colour: Colourless, pale yellow or pale green.

Pleochroism: Coloured grains may display weak pleochroism.

Birefringence: Governed by the substitution of Fe^{3+} for Al. Iron-poor clinozoisites have low birefringence and interference colours of first-order yellow and blue. With increasing iron content and also in thicker grains, interference colours range to brilliant second-order shades. Blue and yellow tints on the margins of the grains are very characteristic. Grains frequently exhibit anomalous blue and yellow interference colours because of strong dispersion.

Extinction: Compositional zoning results in varying extinction angles even within a single grain, and extinction is often incomplete. Crystals elongated on *b* have parallel extinction.

Interference figure: Many specimens, especially those which show incomplete extinction, provide an optic-axis figure with an almost straight isogyre and a few bright orange, red and blue isochromatic curves. Most grains, however, yield poorly defined figures or fail to show any at all. The strong dispersion is rarely visible on detrital species.

Elongation: Either positive or negative. In zoned grains the centre may show a different elongation to that of the outer part.

Distinguishing features: High relief, dominantly prismatic or angular habit, pale colours and often anomalous interference tints aid distinguishing clinozoisite from minerals outside the epidote group. The lack of strong colour and pleochroism and, when obtainable, a positive optical sign help to distinguish clinozoisite from epidote. Zoisite and clinozoisite grains are not always distinguishable from each other. β zoisite especially presents difficulties of identification. α zoisite cleavage fragments usually yield good acute bisectrix figures with strong dispersion and their interference colours are deep indigo blue and dark yellow. Thick non-twinned lawsonite grains resemble clinozoisite, but they are generally basal cleavage fragments, yield centred acute bisectrix figures and normal interference colours. Apatite is uniaxial and has lower birefringence. Vesuvianite shows intense 'Berlin blue', purple or dull yellow interference colours.

Occurrence: Clinozoisite, together with epidote, is a common product of low- to medium-grade metamorphism. They occur in albite–actinolite–epidote–chlorite schists of the greenschist facies, and sometimes in blueschists. Some basic rocks may contain clinozoisite. Epidote, clinozoisite and zoisite are constituents of **saussurite**, which is formed by the alteration of plagioclase.

Grains from: (a) Oligocene Molasse Savoy, France; (b) glacial sand, Norway; (c) Carboniferous, North Sea; (d) Tertiary, North Sea (Mmt 1.662).

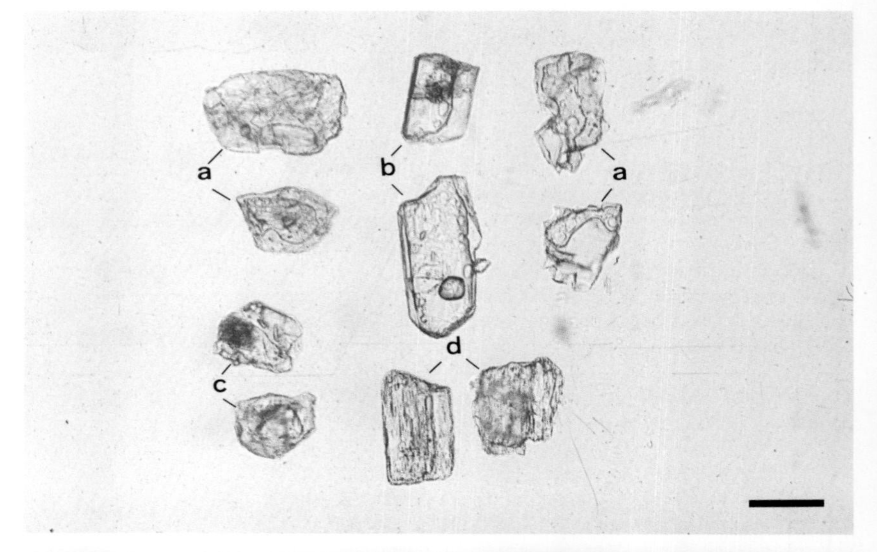

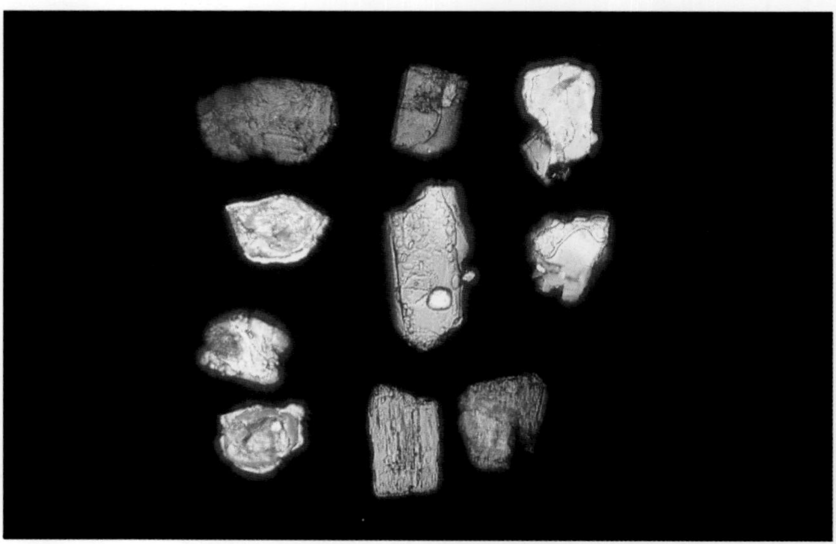

Epidote

$Ca_2Fe^{3+}Al_2O.OH[Si_2O_7][SiO_4]$
monoclinic, biaxial (−)

$n\alpha$	1.715–1.751
$n\beta$	1.725–1.784
$n\gamma$	1.734–1.797
δ	0.015–0.049
Δ	3.38–3.49

Form in sediments: Grains have a fairly high relief and are mostly irregular, angular, equant and sometimes platy. Their corners and indentations may be smoothened by abrasion, but well-rounded or spherical grains are not common. Stumpy prisms or broken euhedral grains are occasionally encountered. Granular aggregates, often embedded in micas, may be frequent in certain formations. Inclusions are fluid globules, opaque particles, rutile and quartz.

Colour: Almost always in shades of green, usually yellowish green or grass green. Grains may exhibit irregular colour distributions.

Pleochroism: Either weak or distinct: α, colourless, pale yellow or pale green; β, greenish yellow; γ, yellowish green.

Birefringence: Moderate, but increases to strong with higher Fe^{3+}. Interference colours are brilliant second- and third-order tints in which green is a dominant shade. Interference colour bands are either concentric or occur in a patchy arrangement enhancing the vivid tints. Thick grains appear uniformly green or whitish green under crossed polars. Pale varieties may exhibit yellow, pale orange and blue polarization colours, similar to those of clinozoisite.

Extinction: Prismatic grains have parallel or nearly parallel extinction.

Interference figure: Centred acute bisectrix figures are rarely seen in epidote grains. Most of them provide optic axis figures with an almost straight isogyre, surrounded by numerous isochromatic curves.

Elongation: Longitudinal sections have negative or occasionally positive elongation.

Distinguishing features: Epidote is diagnosed by high relief, characteristic green colours and pleochroism as well as dominantly irregular morphology. Aggregates can be recognized by their colours and vivid interference tints. The optically negative sign and deeper colours distinguish epidote from clinozoisite, though distinction of these two species is not always certain and they are often combined into one group. Augite is mostly prismatic, has pyroxene cleavages, frequent etch features and it is either non-pleochroic or pleochroic in shades of bluish green. Pumpellyite can be confused with epidote, but it has a lower relief, it is dominantly fibrous, pleochroic in shades of bluish green, and it shows anomalous interference colours. The relief of the green hornblende varieties is lower and they are mostly prismatic. Glauconite occurs as pelletal aggregates or granules, and exhibits aggregate polarization.

Occurrence: Epidote is the index mineral of the albite–

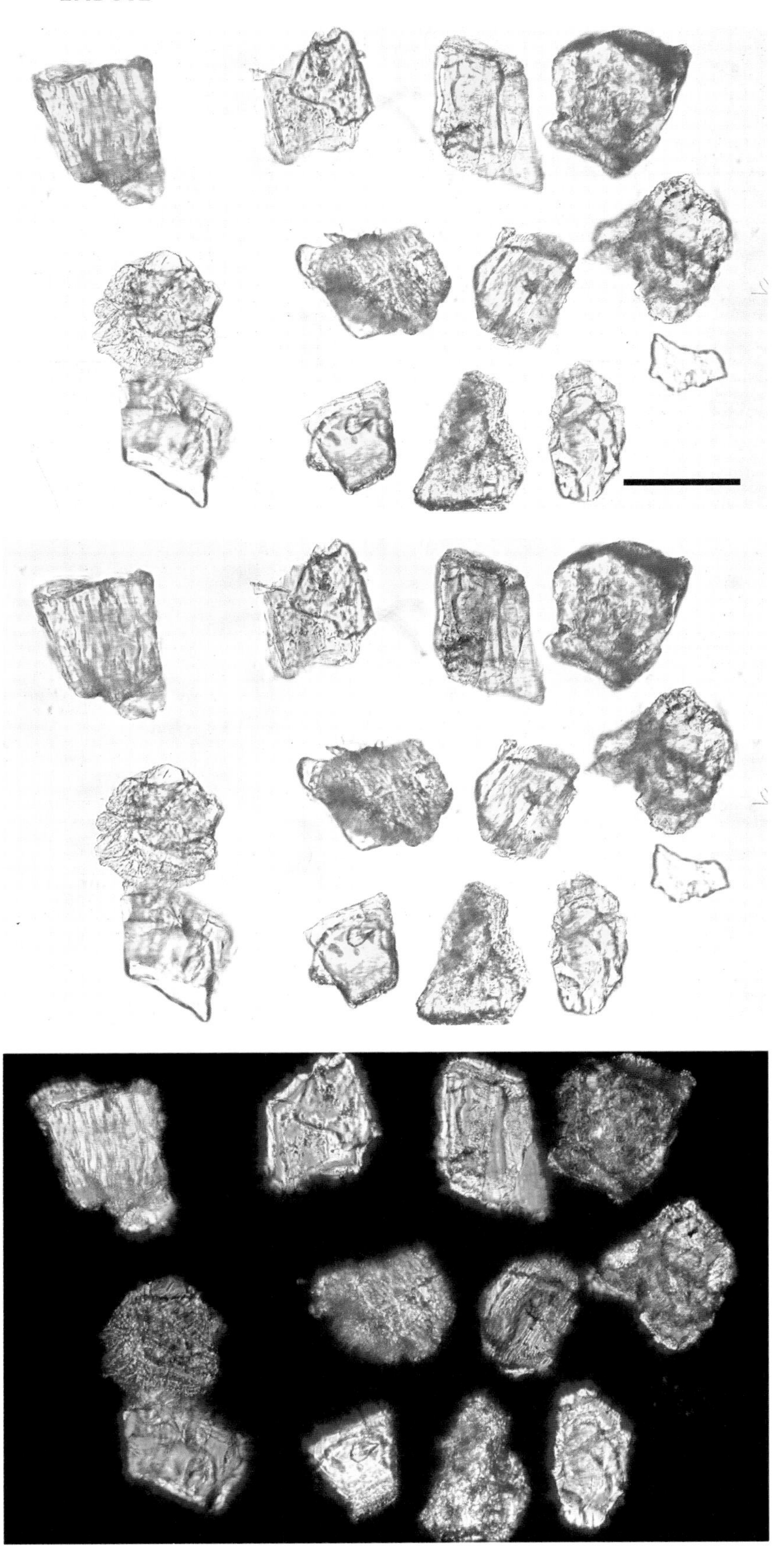

actinolite–epidote–chlorite zones of the greenschist-facies regional metamorphism. It is associated with hornblende in the albite–epidote–amphibolite facies and is also present in contact metamorphic rocks and in hornfelses. Ca-metasomatism results in epidotization. In igneous rocks epidote is more common in the basic types, but may occur in granites. Hydrothermal processes form epidote in cavities, vugs and vesicles in volcanic rocks.

Remarks: Epidote is a widespread detrital mineral, but it is relatively unstable in the diagenetic environment. However, well-sealed formations may contain epidote and one or two grains are occasionally encountered in deeply buried sediments.

Grains from: Carboniferous, North Sea (Mmt 1.662).

Piemontite

$Ca_2(Mn^{3+},Fe^{3+},Al)_2AlO.OH[Si_2O_7][SiO_4]$
monoclinic, biaxial (+)

$n\alpha$	1.732–1.794
$n\beta$	1.750–1.807
$n\gamma$	1.762–1.829
δ	0.025–0.088
Δ	3.45–3.52

Form in sediments: Piemontite occurs most commonly as irregularly shaped angular grains with a fractured, uneven surface. Prismatic fragments, rectangular forms and subrounded abraded grains may also be encountered. Perfect {001} cleavage is rarely detectable. Fluid inclusions, arranged parallel with the *b* axis of the crystals may be present.

Colour: Pale or deep pink, violet, brownish pink or brownish red.

Pleochroism: Strong: α, yellow, orange yellow; β, amethyst, violet, deep red; γ, deep red, brownish red, bright rose pink.

Birefringence: Moderate to strong birefringence results in brilliant second- and third-order interference tints, but these are often masked by intense mineral colour.

Extinction: It is difficult to measure the extinction angle owing to the irregular habit of most grains. Elongated fragments and prisms have parallel extinction.

Interference figure: Prisms laying on (100) provide a nearly centred acute bisectrix figure with a large 2V. Isochromes are few or absent. Irregular fragments and cleavage flakes yield highly eccentric figures. $r>v$ or $r<v$ dispersion is distinct. Piemontite is biaxial positive whereas the negative variety is usually designated as **manganepidote**.

Elongation: Either positive or negative.

Distinguishing features: High relief, intense colour and strong pleochroism suffice for the diagnosis of piemontite. Thulite (a variety of zoisite) has lower refractive indices and is very rare. Highly coloured and pleochroic dumortierite may resemble piemontite, but the prismatic habit, striations and lower relief of the former serves to distinguish it from piemontite. Strongly pleochroic andalusite is likely to be confused with piemontite, but it has lower RI and birefringence, a paler pleochroic scheme and it is optically negative. Hypersthene is pleochroic in shades of pink, pale brown and green and is optically negative.

Occurrence: Piemontite occurs in low-grade regionally metamorphosed schists, gneisses and in vesicles and fissures of acid volcanic rocks. Manganese-rich piemontites appear in manganese deposits of metasomatic or hydrothermal origin.

Grains from: River sand, Isère, France (Mmt 1.662).

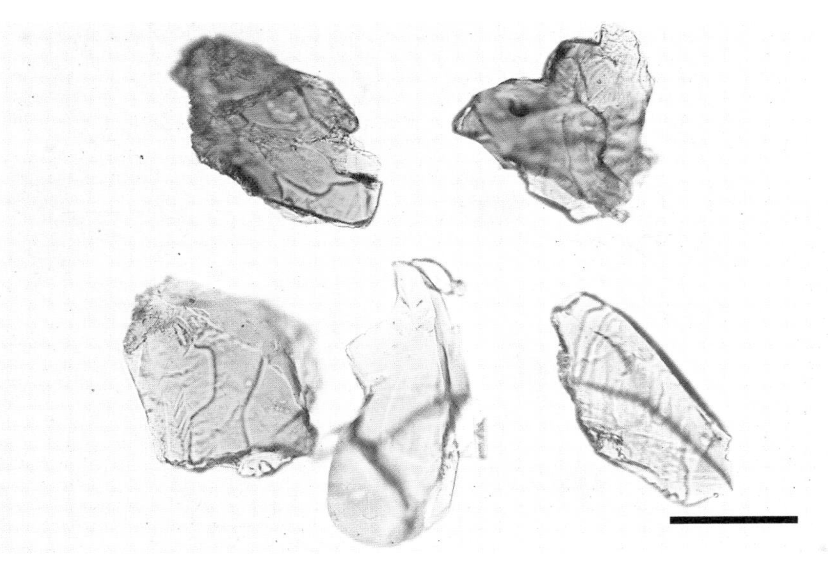

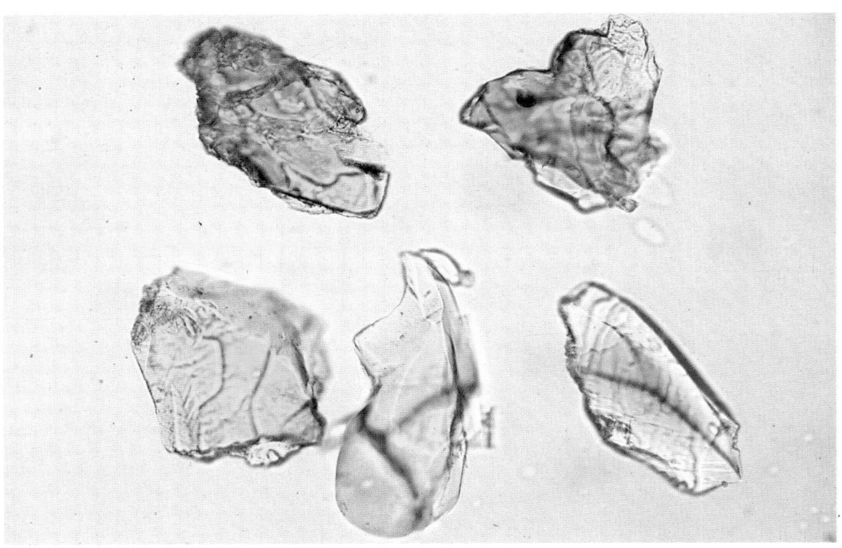

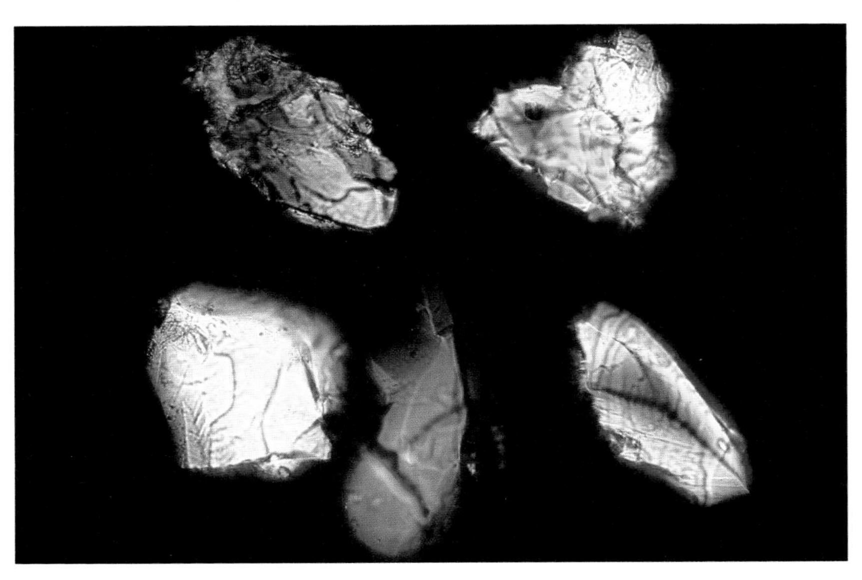

Allanite

$(Ca,Mn,Ce,La,Y,Th)_2(Fe^{2+},Fe^{3+},Ti)(Al,Fe^{3+})_2O.OH$
$[Si_2O_7][SiO_4]$
monoclinic, biaxial (−) (+)

$n\alpha$	1.690–1.791
$n\beta$	1.700–1.815
$n\gamma$	1.706–1.828
δ	0.013–0.036
Δ	3.4–4.2

Form in sediments: Allanite occurs as irregular, angular or somewhat rounded grains, also as fragments of euhedral tablets. Cleavage traces are generally absent, but zoning is common. Grains may be found intergrown with other members of the epidote group.

Colour: Light to dark brown or reddish brown. Irregular colour variations, or grains displaying a dark core or lighter margins are not uncommon.

Pleochroism: Allanite is strongly pleochroic: α, light brown, reddish brown; β, brownish yellow, or brown; γ, greenish brown, very dark brown.

Birefringence: Moderate and interference colours are bright second-order tints, but are often masked by the deep colour of the mineral.

Extinction: Elongated fragments give parallel extinction.

Interference figure: The interference figure is usually obscured by strong mineral colour and absorption.

Elongation: Both positive and negative elongations have been reported.

Distinguishing features: Allanite strongly resembles the brown hornblende varieties, which may cause confusion. However, in sediments the latter occur as prismatic fragments, and usually show amphibole cleavages and inclined extinction. The relief of brown tourmaline is lower and is uniaxial. Dark brown biotites are platy, have lower RI and weaker pleochroism.

Occurrence: Allanite is an accessory mineral of granite, granodiorite and syenite. It also forms in some pegmatites, schists and gneisses, and rarely in volcanic rocks.

Remarks: Radioactive components (U and Th) are usually present in allanite. Radioactive bombardment destroys the crystal structure, the mineral then reaches a metamict state and becomes isotropic.

Grains from: Upper Eocene tuff horizon, Possagno, Italy (Mmt 1.662).

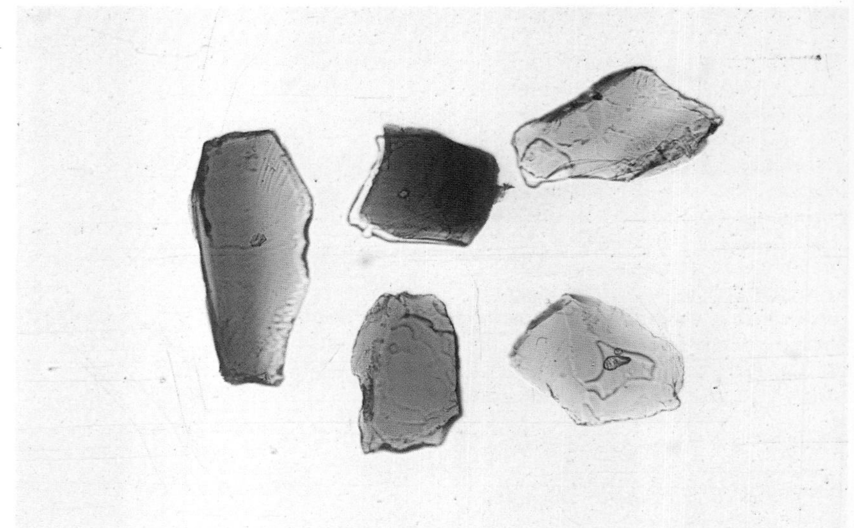

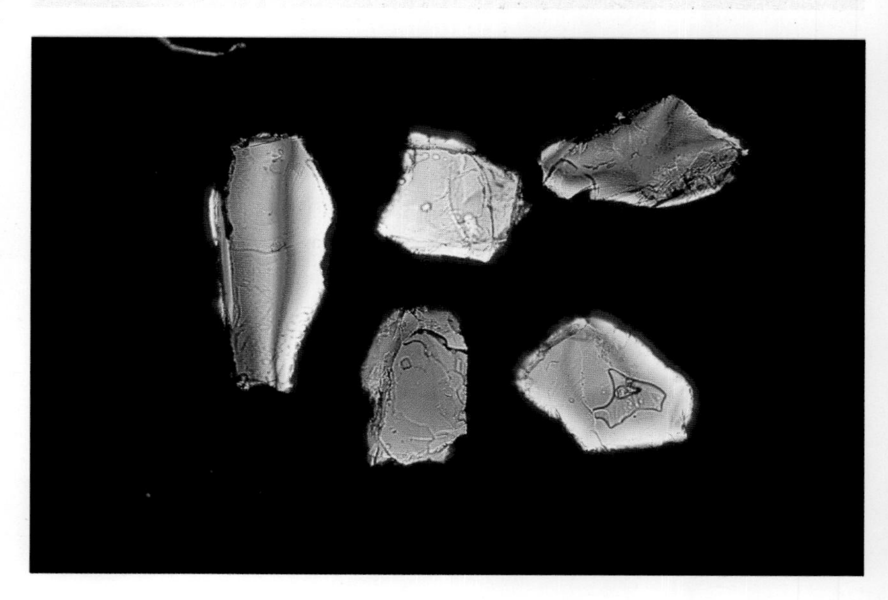

Lawsonite

$CaAl_2[Si_2O_7](OH)_2H_2O$
orthorhombic, biaxial (+)

$n\alpha$	1.665
$n\beta$	1.672–1.676
$n\gamma$	1.684–1.686
δ	0.020
Δ	3.05–3.12

Form in sediments: The shape of detrital lawsonite grains is determined by the perfect {100} cleavage and they occur as platy angular, equidimensional, irregular or rhombic fragments. The edges may be rounded. Sometimes a complete euhedral crystal appears (upper, a). Fluid inclusions, (mainly CO_2) are not uncommon. Twinning is frequent and is either simple lamellar or multiple in two directions (parquet twinning; lower). The thin twin lamellae are sometimes bent.

Colour: Lawsonite is colourless and has a high lustre. Thicker grains may exhibit a bluish tinge, or are pale yellow.

Pleochroism: α, pale blue; β, bluish green; γ, pale yellow, noticeable only on very thick grains.

Birefringence: Lawsonite has a weak birefringence and the interference colours are first-order white, grey or pale yellow. Thick grains display intensive second-order orange, blue and red colours.

Extinction: {100} cleavage plates show symmetrical extinction. That of longitudinal fragments is parallel.

Interference figure: {100} cleavage flakes provide a well-centred acute bisectrix figure with bright isochromatic curves, usually discernible only in thicker grains, and a clear positive sign. $r>v$ dispersion is detectable.

Elongation: Positive in crystals elongated on *y* and in rhombic plates in the direction of the long diagonal.

Distinguishing features: The platy, generally angular form, high lustre and twinning are diagnostic of lawsonite. Untwinned, thicker grains are most likely to be mistaken for clinozoisite, especially for those which fail to show anomalous interference colours. However, clinozoisite has inclined extinction and usually yields an optic axis figure with a straight isogyre, whereas lawsonite displays a clear acute bisectrix figure. The anomalous interference colours of zoisite assist in its distinction from lawsonite. Andalusite is biaxial negative and pleochroic in shades of pink and red. Topaz occurs mainly as irregular, thick, somewhat rounded or chipped grains and has smaller 2V. Twinned plagioclase grains, included accidentally in the heavy fraction, may be confused with lawsonite, but their relief is considerably lower.

Occurrence: Lawsonite is a diagnostic mineral of the high-pressure metamorphism and forms in blueschists, metaophiolites and in some eclogites.

Remarks: Blue amphibole in the heavy mineral fraction may signal the probability that lawsonite is present.

Grains from: Upper: (a) Lower Miocene Molasse, bore-

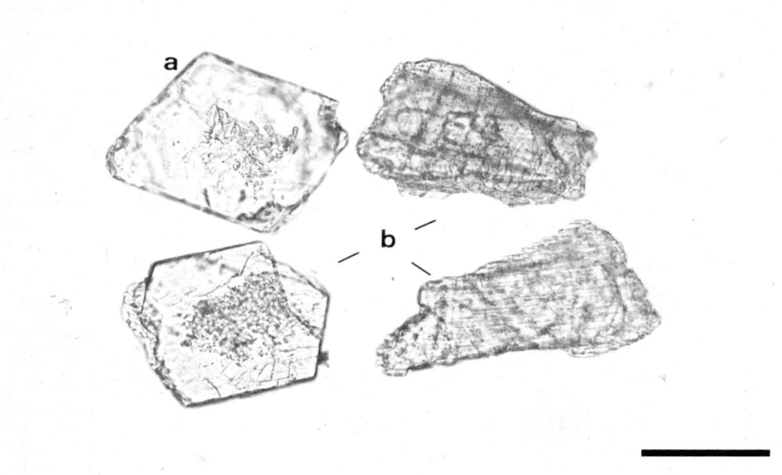

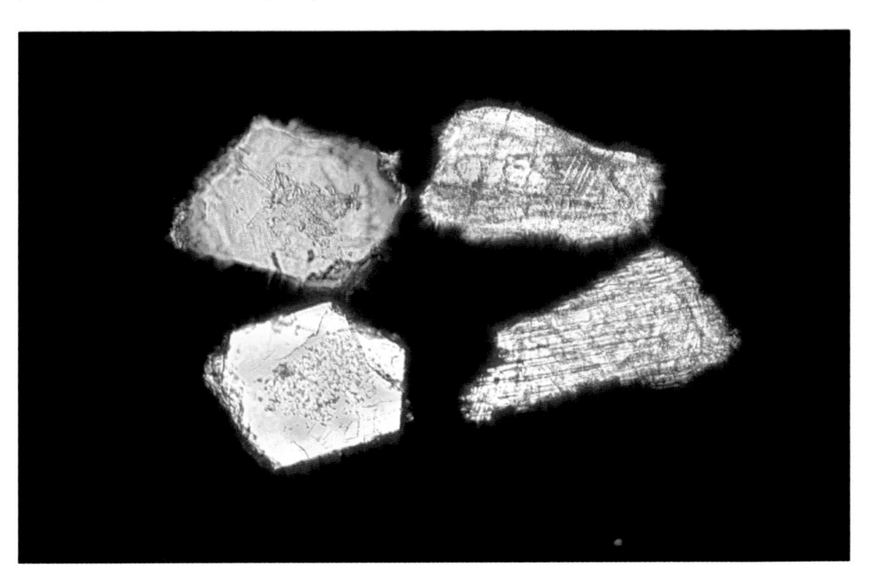

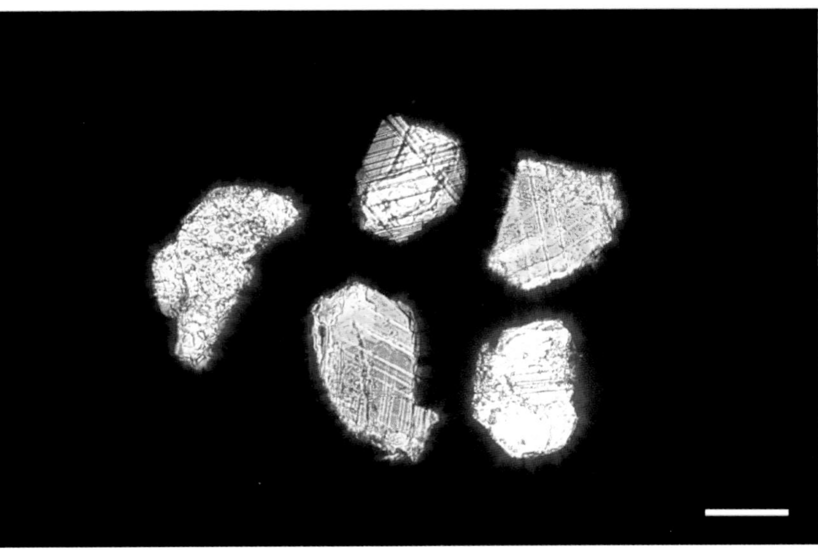

hole Savoy-104, 1242 m France; (b) Oligocene, Barrême Basin, France (Mmt 1.582).
Lower: Oligocene Molasse, Savoy, France (Mmt 1.582).

Pumpellyite

$Ca_4(Mg,Fe^{2+},Mn)(Al,Fe^{3+},Ti)_5O(OH)_3[Si_2O_7]_2[SiO_4]_2.2H_2O$
monoclinic, biaxial (+)

$n\alpha$	1.674–1.748
$n\beta$	1.675–1.754
$n\gamma$	1.688–1.764
δ	0.014–0.022
Δ	3.18–3.23

Form in sediments: Pumpellyite occurs as radial fibres arranged either within a fan-shape form or in parallel to subparallel growth, also as clusters of rosette-like needles, bladed crystals and dense aggregates. Well-formed prisms or angular grains are very rare.

Colour: Various shades of green colours are dominant which range from pale yellowish green to deep bluish green. Some grains show a patchy colour distribution. Brown, yellowish-brown or colourless grains are infrequent.

Pleochroism: Pumpellyite is highly pleochroic with: $\alpha = \gamma$, colourless, pale yellow, pale greenish yellow; β, light green, deep bluish green. Crystals are usually elongated parallel to y (y = β) and the deep bluish-green or green β maximum absorption is well displayed on many detrital grains.

Birefringence: Both the thickness of the grains and the orientation of the fibres influence the intensity of the interference colours. These may be first-order white or yellow on thin grains. Because of strong dispersion of the optic axes, most pumpellyite exhibits anomalous brown to yellowish and greyish-blue interference colours. Thicker grains show interference colours up to second-order yellow, orange and blue.

Extinction: Within the clusters of fibres or laths, the extinction of the individual crystals is approximately parallel.

Interference figure: Prismatic varieties may display a centred acute bisectrix figure on a bluish field and strong $v>r$ dispersion. Broader fibres yield off-centre interference figures.

Elongation: Commonly negative, occasionally positive.

Distinguishing features: The fibrous habit, often radially arranged fibres or needles, characteristic pleochroism, anomalous interference colours and generally negative elongation are reliable properties for the diagnosis of pumpellyite. The moderately high relief serves to distinguish it from strongly pleochroic chlorite. Pumpellyite is most likely to be mistaken for the epidote minerals. However, its fibrous habit, lower RI distinct pleochroic scheme (especially its bluish shade) and the anomalous interference colours assist in the distinguishing pumpellyite from the former group. Fibrous amphiboles show higher-order interference colours, larger extinction angle and positive elongation. Some pumpellyite aggregates resemble glauconite, but the relief of the latter is lower. Prismatic pumpellyites are similar to the green andalusite variety, viridine, but both are very rare in sediments. More details on detrital pumpellyite can be found in Mange-Rajetzky & Oberhänsli (1986).

Occurrence: Pumpellyite is a characteristic mineral of

blueschists and greenschists and is the index mineral of the low-grade prehnite–pumpellyite facies. It is common in metamorphosed intrusive rocks and may be formed by hydrothermal action in amygdales and in veins. Amygdales in basalt, andesite and spilitic basalt may also contain pumpellyite. It has been reported from meta-greywackes and siliceous gneisses.

Grains from: Lower Miocene Molasse, Yverdon, Switzerland (Mmt 1.582).

Tourmaline group

General formula:
$Na(Mg,Fe,Mn,Li,Al)_3Al_6[Si_6O_{18}](BO_3)_3(OH,F)_4$
trigonal, uniaxial (−)

The tourmaline group comprises three principal compositional varieties:

magnesian tourmaline or dravite

$n\omega$	1.635–1.661
$n\varepsilon$	1.610–1.632
δ	0.021–0.036
Δ	3.03–3.15

iron tourmaline or schorl

$n\omega$	1.655–1.675
$n\varepsilon$	1.625–1.650
δ	0.025–0.035
Δ	3.10–3.25

lithium tourmaline or elbaite

$n\omega$	1.640–1.655
$n\varepsilon$	1.615–1.630
δ	0.017–0.024
Δ	3.03–3.10

Dravite and schorl, as well as schorl and elbaite, form a continuous solid-solution series, but there is an immiscibility gap between elbaite and dravite.

Form in sediments: Three inherent basic habits give rise to extremely variable detrital morphology. These are: short slender prisms with terminations at one, or rarely, at both ends (upper; a), curved near-triangular or six-sided basal sections, produced by closely spaced parting (upper; b and c) and columnar to acicular aggregates or radiating needles (upper; d). The latter two are far less common than the former. During sedimentary processes the prisms fracture, producing prismatic fragments and irregular, angular or equidimensional particles, which in turn attain various degrees of rounding. Ultimately grains become egg-shaped, long ellipsoidal or spherical (second page upper). Deposits with mixed provenance may contain a large variety of tourmaline morphology. Cleavages of tourmaline are very poor, but grains sometimes exhibit striations. Inclusions are common and can indicate particular parageneses, hence they are useful indices of provenance. Inclusions are: rutile, zircon, magnetite, cassiterite, topaz, fluorite, quartz, apatite, titanite, as well as vacuoles, fluid globules and carbonaceous impurities. Secondary growths (overgrowths, second page lower) have been frequently reported on tourmaline grains. These form either authigenically during diagenesis and are in optical continuity with the host grain or, as demonstrated by Henry & Guidotti (1985), develop on the detrital tourmalines of sediments while undergoing metamorphisms and are in optical discontinuity with the detrital grain. The latter may experience rounding after subsequent reworking (second page lower; b). The mode and size of authigenic overgrowths vary from minute protrusions to extreme types. The overgrowth forms in the direction of the c axis and this particular polarity is best seen on prismatic grains where they appear at one end (at the antilogous pole according to Alty 1933). They are usually parallel intergrowths of small individual prisms of varying lengths, which are in optical continuity with the detrital grain,

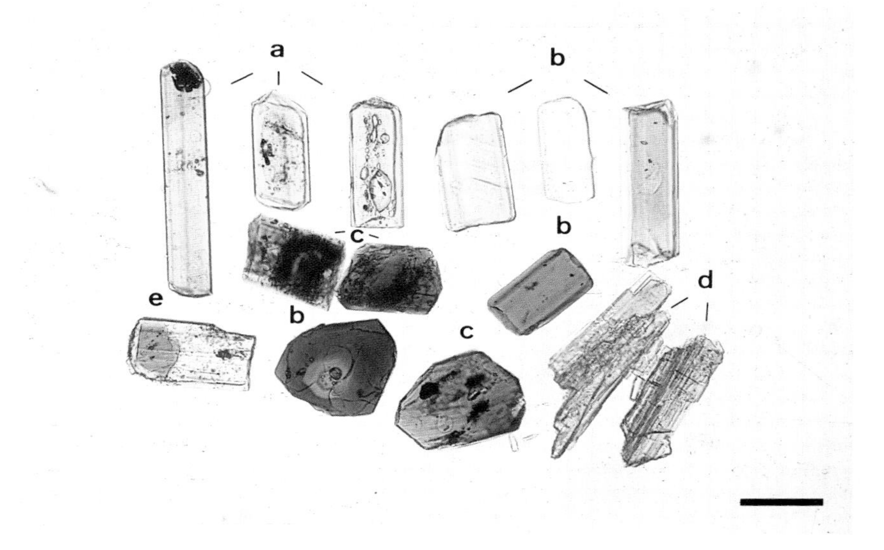

but their colour is usually pale green or blue, but sometimes they lack colour. The overgrowths are usually welded to the host grain by small 'roots'. They frequently enclose minute impurities. Certain features resemble overgrowths and these need to be recognized in order to avoid confusion. Grains having discontinuous optical zoning may appear similar to those with overgrowths, but distinction can be made by the lack of 'roots' on the former. Fracturing and thinning of the grains at one end, or sometimes cementing material adhering to the grain, can be mistaken for an overgrowth, but the latter is in optical discontinuity with the tourmaline grain. The detrital nature of the nucleus can be ascertained by examining it under high magnification and this usually reveals an abraded surface.

Colour: Tourmaline displays a wide range of colours and these are, in general, indications of composition. Iron-bearing tourmalines are very dark (almost opaque) or deep blue, elbaites have light or deep blue (indicolite) and pink (rubellite) shades, and dravite is dark brown, yellow or almost colourless. Unusual tints are produced by various metallic ions. Colour zoning is frequent. This is visible either as zoning along the *c* axis or as colour bands on the basal pinacoid. Varicoloured tourmalines are not uncommon. In this instance the colour is patchy or one portion of the grain displays strikingly different shades to the other.

Pleochroism: Tourmalines, especially the iron-bearing varieties, have strong and distinctive pleochroism. Maximum absorption appears when the long axis is lying perpendicular to the vibration direction of the polarizer. In rare cases, fragments of the basal parting show reversed maximum absorption. Iron tourmalines show pleochroism either from yellow brown to pale yellow or from deep blue, dark green to pale yellow or colourless. The pleochroism of magnesian tourmalines is in shades of pale yellow, lithium tourmalines are colourless in the ε direction and have paler tints of their natural colour (i.e. light blue, green or pink) in the direction of maximum absorption. Basal partings or grains oriented according to the basal pinacoid usually lack pleochroism.

Birefringence: Tourmaline has strong birefringence and high-order interference tints, though these are usually obscured by strong mineral colour. Numerous narrow polarization colour bands are best seen in pale prismatic grains. Basal sections have dull interference tints and some appear almost isotropic.

Extinction: Prisms and fragments elongated on *c* have parallel extinction.

Interference figure: Tourmaline is uniaxial, but prisms, fragments and rounded grains generally yield biaxial figures observed as shadows of two isogyres leaving the field rapidly on stage rotation (similar to biaxial flash figures). Good centred uniaxial figures are exhibited by basal sections.

Elongation: Negative, except for some fragments of the basal parting which may have positive elongation.

Distinguishing features: Tourmaline is one of the most common and easily recognizable detrital minerals. It is diagnosed by moderate relief, highly variable morphology and colours, as well as marked pleochroism, negative elongation and lack of cleavages. Colourless grains may be mistaken for

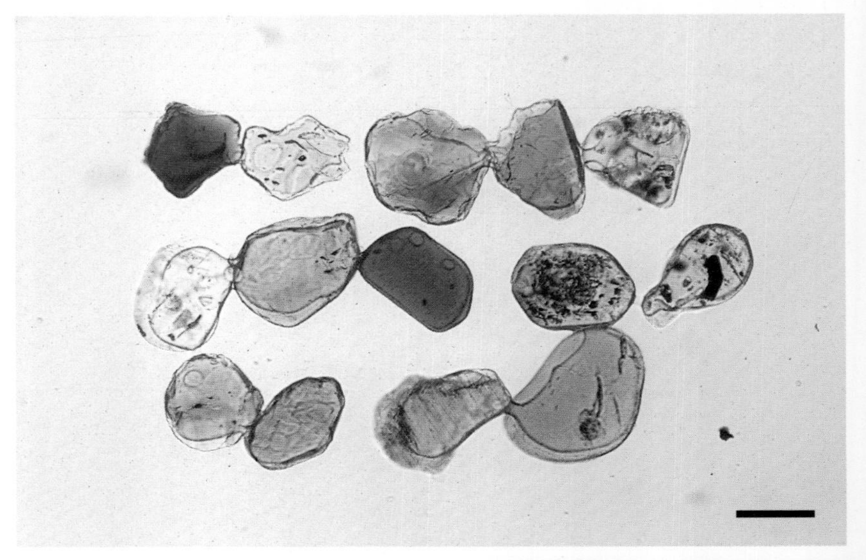

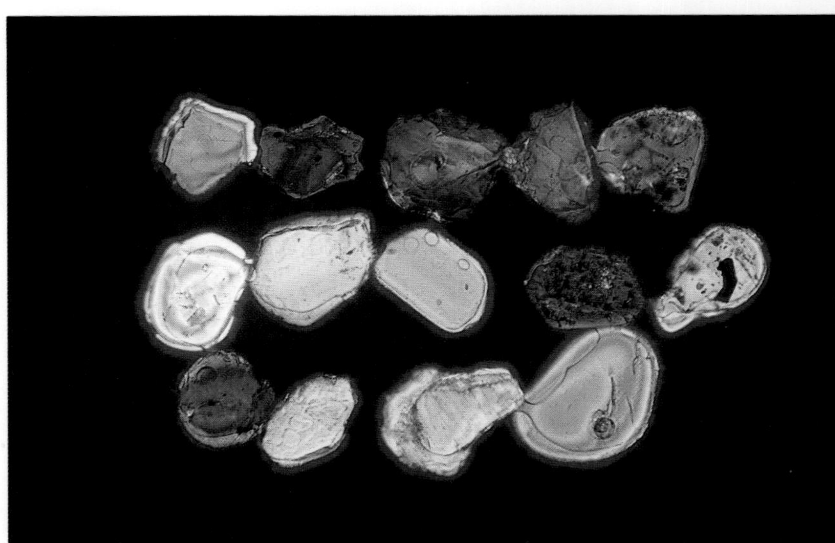

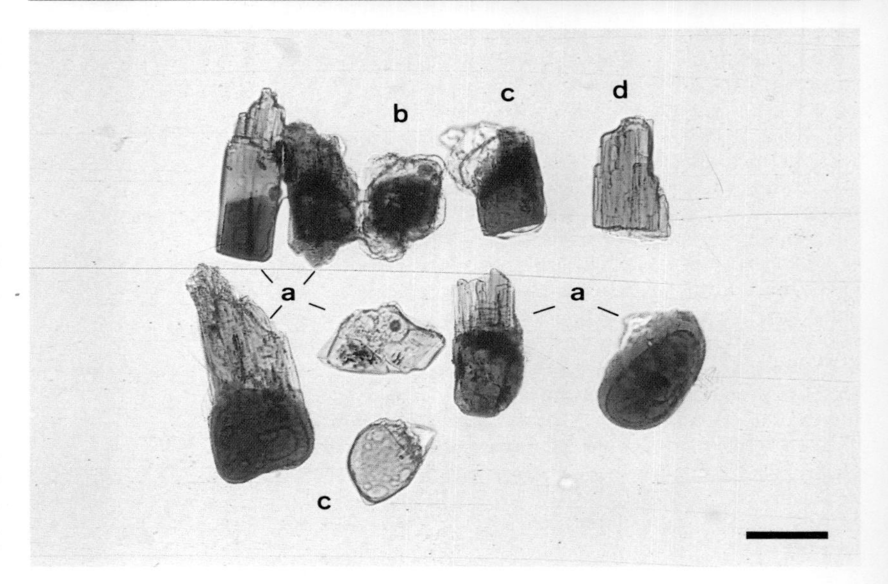

andalusite, topaz or sillimanite, but a clear uniaxial figure on their basal plane assists in diagnosis. Very dark grains can easily be overlooked as opaques. In this case insertion of the accessory condenser usually reveals colours at the thinner ends, or rolling the grains onto another face may yield lighter colours. Green prisms are most likely to be mistaken for aegirine or for certain amphiboles. The blue varieties resemble alkali amphiboles or dumortierite and the brown-coloured ones resemble brown hornblende. Aegirine has a higher relief, weaker pleochroism, stronger interference colours and shows cleavages. The reversed absorption of amphiboles and dumortierite, the positive elongation of most amphiboles (except some alkali amphiboles), as well as the good amphibole cleavages, aid distinction. Hypersthene has higher refractive indices and is length slow. Some yellowish-brownish angular fragments of tourmaline resemble staurolite, but the latter has a higher relief and lower birefringence. The diagnosis of well-rounded tourmalines presents no difficulties, as other coloured minerals (apart from rutile) seldom attain such extensive rounding. In aeolian sediments amphiboles and pyroxenes may be similarly rounded, but cleavage traces of the former and the higher relief of the latter help to distinguish them.

Occurrence: Tourmaline crystallizes in granites, granite pegmatites, in pneumatolitic veins and in contact- or regionally metamorphosed rocks. In schists, gneisses and phyllites it may form by metasomatism or occur as recrystallized detrital grains. Tourmalines of the schorl–elbaite series are found in granitoid rocks. Lithium tourmalines, often with high variation of colours and composition, are present in some pegmatites and late-stage granitic veins. Magnesian tourmalines occur in some metamorphic schists, in metasomatic rocks and in certain basic igneous rocks.

Remarks: Tourmalines are widespread in all types of detrital sediments and are ultrastable both mechanically and chemically. Polycyclic tourmaline grains eroded from pre-existing siliciclastic deposits are well to very well rounded and are associated with equally well-rounded zircons and rutiles. The mineralogical maturity of a sediment can be expressed by the zircon–tourmaline–rutile (ZTR) index (Hubert 1962). The study of varietal types, which aims at distinguishing various morphology, colours, zoning and inclusions of tourmaline grains, can reveal significant trends. Characteristic tourmaline types may prove important for correlating certain sand bodies, tracing sediment source and dispersal (p.19). A condensed review of the tourmaline group in sediments was presented by Krynine (1946).

Grains from: First page upper: (a) Tertiary, North Sea; (b) Jurassic, North Sea; (c) Aquitanian, Nerthe-Chain, south of France; (d) Carboniferous, borehole Weiach, 1431 m, Switzerland; (e) Oligocene Molasse, borehole Weiach, 51 m, Switzerland (Mmt 1.582).
Second page upper: Polycyclic grains from various mature sediments (Mmt 1.582).
Second page lower: Grains are from the Lower Cretaceous, northern Tunisia: (a) detrital grains with authigenic overgrowths; (b) grain with rounded overgrowth; (c) authigenic overgrowth intergrown with quartz; (d) detached authigenic overgrowth (Mmt 1.582).

Axinite

$(Ca,Mn,Fe^{2+})_3Al_2BO_3[Si_4O_{12}]OH$
triclinic, biaxial (−)

$n\alpha$	1.674–1.693
$n\beta$	1.681–1.701
$n\gamma$	1.684–1.704
δ	0.009–0.011
Δ	3.26–3.36

Form in sediments: Axinite appears as irregular, equidimensional, rarely diamond-shaped grains. Their corners are often rounded and the crystal faces frequently show small indentations, etch patterns or small conchoidal fractures. The distinct {100} cleavage may be detectable. Axinite crystals are commonly composed of wedge-shaped sectors which may be recognized in some grains. Opaque and accessory mineral inclusions are frequent.

Colour: Pale brown and dull yellowish brown, usually with a purplish hue, are the commonest shades but some grains are colourless, or violet, and those of the manganoan variety display yellow, orange and red colours. Colour zoning may be present.

Pleochroism: Detrital axinites are moderately pleochroic with a tinge of slight pink, violet or purplish: α, yellow to light purplish brown; β, bluish or violet brown; γ, pale brown, pale violet, or brownish green.

Birefringence: Weak and the interference colours are grey, white or first-order yellow. On thick grains the colours range up to second-order orange or green, but they are usually masked by the natural colour of the mineral.

Extinction: Inclined to cleavage traces, but grains often fail to extinguish in any position.

Interference figure: The majority of axinite grains give an off-centre optic axis figure with a nearly straight isogyre. The thicker grains show several isochromes and strong v>r dispersion.

Elongation: Longitudinal sections have positive elongation.

Distinguishing features: High relief coupled with low birefringence, dull colours and their pinkish or purplish tinge, as well as moderate pleochroism, are characteristic. The relief of brown tourmaline is lower; it is strongly pleochroic and uniaxial. The pleochroism of allanite is stronger. Staurolite has brighter colours and more intensive interference tints. The large extinction angle of titanaugite, together with its high-order interference colours, serve to distinguish it from axinite.

Occurrence: Axinite is a characteristic mineral of contact metamorphic aureoles between carbonate rocks and garnitic intrusives. It occurs also in veins and cavities in granites, mafic intrusives and basalt.

Grains from: Beach sand, St Ives, Cornwall, England (Mmt 1.662).

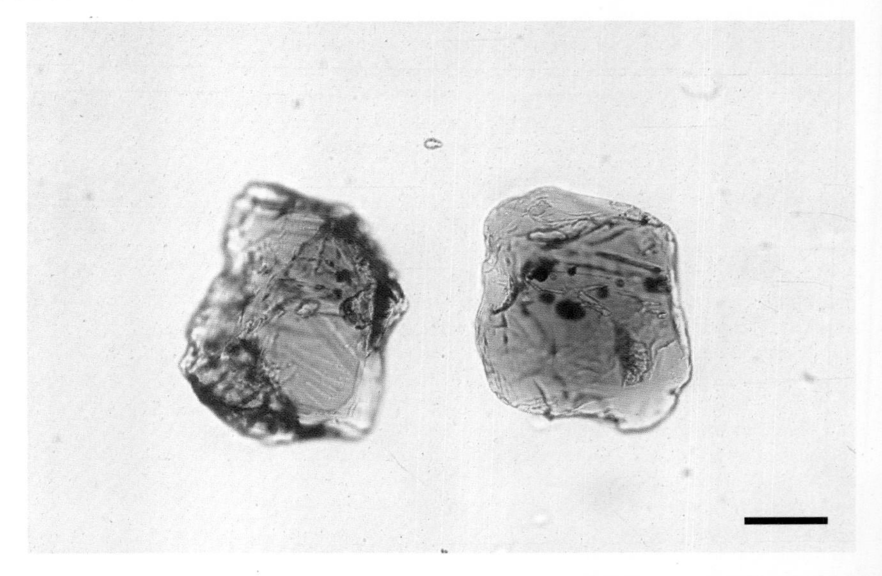

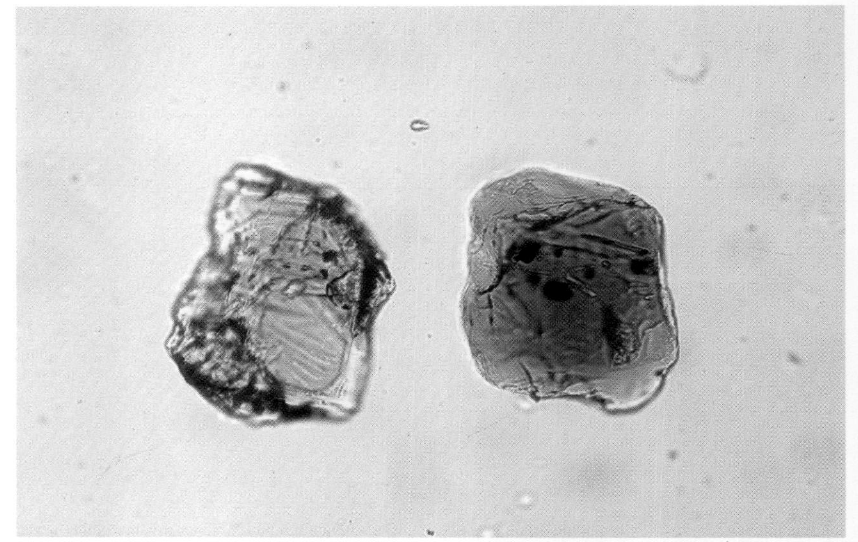

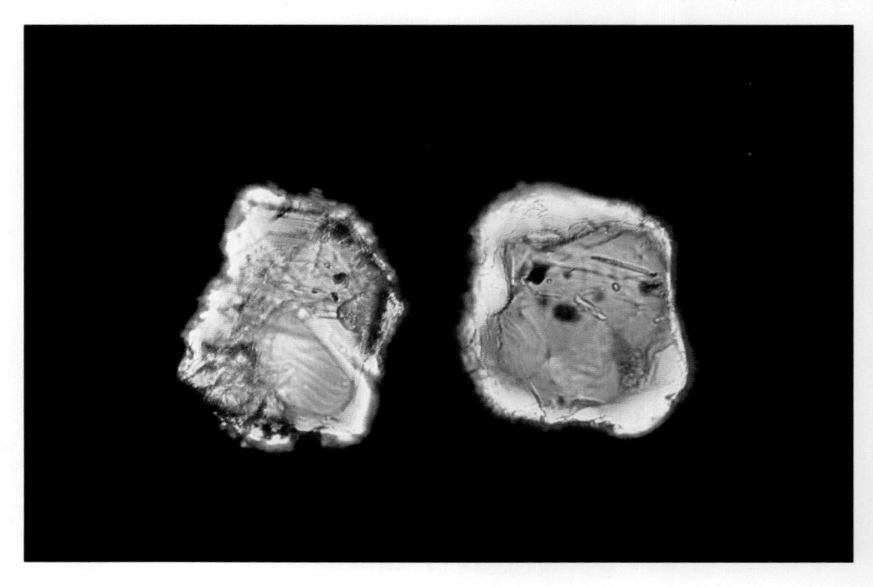

PYROXENE GROUP

General considerations

Pyroxenes are one of the most important rock-forming minerals, but of the many members of the pyroxene group only a few occur commonly as detrital grains. Pyroxenes are chemically unstable in sediments and they dissolve at an early stage of diagenesis. Therefore, they are found only in either well-sealed ancient sediments or in younger deposits, but in both cases they are valuable indicators of provenance.

The chemistry of a particular pyroxene species provides the basis of accurate identification. Chemical data plotted in the triangular diagram of Poldervaart & Hess (1951), proposed for the $CaMgSi_2O_6$–$CaFeSi_2O_6$–Mg_2Si_2O–$Fe_2Si_2O_6$ system, will enable any pyroxene species to be named (see Deer *et al.* 1978, p. 3) thus providing useful clues to provenance.

Within the pyroxene group two principal subdivisions exist: **orthopyroxenes** with orthorhombic and **clinopyroxenes** with monoclinic symmetry. Based on crystal chemistry, pyroxenes are classified as follows (Deer *et al.* 1978):

Magnesium–iron pyroxenes
Orthopyroxenes (enstatite–orthoferrosilite)
$(Mg,Fe^{2+})_2Si_2O_6$
Pigeonite $(Mg,Fe^{2+},Ca)(Mg,Fe^{2+})Si_2O_6$

Calcium pyroxenes
Diopside-Hedenbergite $Ca(Mg,Fe^{2+})Si_2O_6$
Augite $(Ca,Mg,Fe^{2+}Al)_2(Si,Al)_2O_6$
Fassaite $Ca(Mg,Fe^{2+},Fe^{3+},Al)(Si,Al)_2O_6$
Johannsenite $CaMnSi_2O_6$

Calcium-sodium pyroxenes
Omphacite (aegirine–augite)
$(Ca,Na)(Mg,Fe^{2+},Fe^{3+},Al)Si_2O_6$

Sodium pyroxenes
Jadeite $NaAlSi_2O_6$
Aegirine $NaFe^{3+}Si_2O_6$
Ureyite $NaCrSi_2O_6$

Lithium pyroxenes
Spodumene $LiAlSi_2O_6$

Of the above the following species have been reported from sediments:
Orthopyroxenes of the solid solution series **enstatite–orthoferrosilite** with intermediate members bronzite, hypersthene, ferrohypersthene and eulite. In cases where grains are diagnosed only by their optical properties, the colourless and optically positive pyroxenes are identified as enstatite and the coloured, pleochroic and optically negative species as hypersthene.

Calcium pyroxenes: Certain menbers of the diopside–salite–ferrosalite–hedenbergite solid-solution series, and especially augite minerals, are present in many detrital sediments.

Calcium–sodium pyroxenes: Aegirine–augite may be important locally.

Sodium pyroxenes: In isolated occurrences aegirine has been reported both as detrital and as authigenic.

Lithium pyroxene spodumene has been reported from sediments (Milner 1962), but it has a restricted paragenesis and it generally forms as large crystals. Therefore, the chances of finding it in amounts sufficient for positive identification are rather remote. Its description is omitted from the present book.

Enstatite

$Mg_2[Si_2O_6]$
orthorhombic, biaxial (+)

$n\alpha$	1.650–1.662
$n\beta$	1.653–1.671
$n\gamma$	1.658–1.680
δ	0.007–0.011
Δ	3.21–3.96

Form in sediments: Grains are dominantly long or short stumpy prisms, irregularly terminated prismatic fragments, cleavage fragments or more rarely anhedral debris. The corners of some grains may be rounded. Iron-bearing enstatites often have a yellow stain because of oxidation of Fe^{2+} to Fe^{3+}. Prismatic cleavages, solution grooves, 'hacksaw terminations' or serpentinization (known as **bastite**) are frequently exhibited. Lamellar structure is fairly common, which is attributed to exsolution of a clinopyroxene phase or to translation gliding, sometimes twinning. This structure is best observed under crossed polarizers when the main part of the mineral is in extinction position. Inclusions may be magnetite, apatite and zircon.

Colour: Enstatite is essentially colourless. Some large thicker grains may have a pale green tinge.

Pleochroism: Not visible on detrital species.

Birefringence: Depends on the ferrous iron content and is weak to moderate. Interference colours of thinner grains are first-order white or yellow, but thick grains exhibit a wide range of polarization colours of second-order green, yellow, mauve and blue.

Extinction: Of prisms and longitudinal cleavage fragments is parallel.

Interference figure: Acute bisectrix figures are provided by basal sections. Prismatic faces usually yield flash or eccentric optic axis figures.

Elongation: Positive.

Distinguishing features: Enstatite is diagnosed by moderate relief, lack of colour, prismatic morphology, cleavages, frequent lamellar structure and parallel extinction. Chromian spinel, serpentine and olivine in the heavy mineral fraction may also indicate the presence of enstatite. It is distinguished from hypersthene by the colour and pleochroism of the latter and from clinopyroxenes by its parallel extinction. Diallage has a platy habit, exhibits parting and a characteristic optic axis figure. The RI of anthopyllitic amphibole is lower and is usually optically negative. Sillimanite has stronger birefringence and mostly a long slender prismatic morphology.

Occurrence: Magnesium-rich orthopyroxenes are the chief constituents of ultramafic rocks such as peridotite, pyroxenite, harzburgite, lherzolite, serpentine, etc. They appear in some meta-mafic rocks and rarely in granulites.

Grains from: Lower right: beach sand, Kynance Cove, Cornwall, England; Other grains are from a beach sand, southern Turkey (Mmt 1.582).

Hypersthene

$(Fe^{2+},Mg)_2[Si_2O_6]$
orthorhombic, biaxial (−)

$n\alpha$	1.669–1.755
$n\beta$	1.674–1.763
$n\gamma$	1.680–1.773
δ	0.018–0.020
Δ	3.21–3.96

Form in sediments: Hypersthenes of intrusive and plutonic rocks are irregularly shaped fragments of larger crystals, cleavage pieces or short stubby prisms (b). Some grains may be serpentinized (**bastite**). Hypersthenes sourced by effusive rocks are long slender or short prisms, broken euhedral crystals and those of pyroclastic origin often appear as complete euhedral crystals (a) and are sometimes surrounded by frothy volcanic glass or matrix. Grains are either sharp, angular or their corners may be rounded. Etch patterns frequently develop on all types and are seen as hacksaw terminations, dissolution voids, grooves, or sometimes mamillae (c). Inclusions are common, especially in the effusive types. These are opaque particles, zircon, apatite, feldspar, numerous fluid globules or gas. Minute plates and rods of ilmenite, brookite and titaniferous magnetite inclusions, often appearing in parallel arrangements, are called **Schiller-structure** and are attributed to material exsolved with slow cooling.

Colour: Shades of pink, pale reddish brown and green.

Pleochroism: Distinct and shows varying intensities. It may be very strong in grains of volcanic origin: α, pale reddish brown, purple-violet, pink; β, pale greenish brown, pale reddish yellow, pale brown, yellow; γ, pale green, smoky green, grey–green, green.

Birefringence: Moderate to strong. Thin prisms and cleavage fragments as well as {010} faces have first-order white, yellow and orange interference colours without colour bands. Thicker grains exhibit vivid second- and third-order polarization colours with numerous colour bands, arranged parallel with the elongation of the prisms.

Extinction: Of prisms and cleavage fragments is parallel. Occasional basal sections have symmetrical extinction.

Interference figure: {010} faces and parting plates, which often have a broader form and lower-order interference colours, yield centred or nearly centred acute bisectrix figures with large 2V. Isogyres are well defined in a yellow or white field. Isochromes are either absent or some appear in the periphery of the field. Other sections show eccentric optic axis or flash figures. v>r dispersion is moderate to strong.

Elongation: Positive.

Distinguishing features: The most diagnostic feature of hypersthene is the characteristic pleochroism. Together with the optically negative sign, it serves to distinguish hypersthene from enstatite and diallage. Clinopyroxenes have a large extinction angle. Andalusite (pleochroic in shades of pink and red) may be mistaken for hypersthene, but the former has a lower relief, is length fast and has poorly developed cleavages.

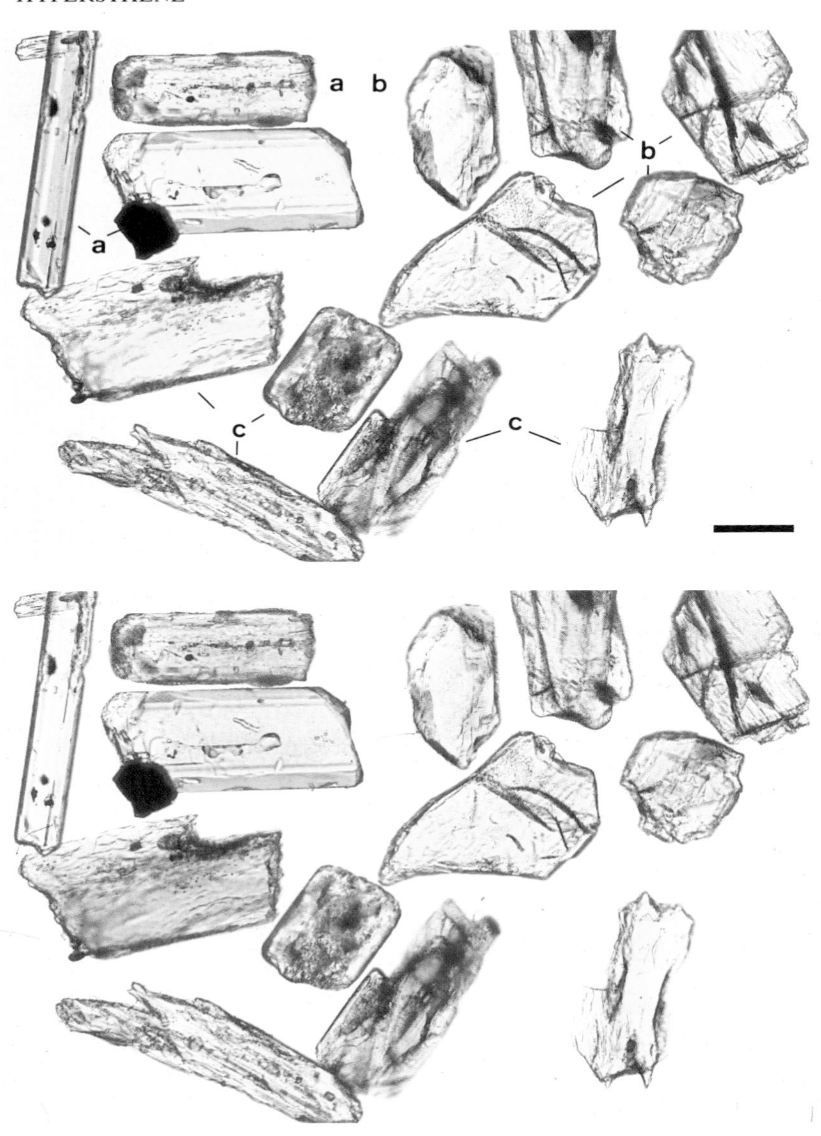

Piemontite has higher refractive indices and is optically positive. Tourmaline is uniaxial and is length fast, whereas the relief of hornblende is lower, it has inclined extinction and more enhanced cleavages.

Occurrence: Hypersthene is common in both extrusive and intrusive basic to intermediate igneous rocks, such as gabbros, norites, basalt, andesite and dacite. It appears in some granites and syenites. Hypersthene has been reported from metamorphic rocks which include charnokites, hypersthene granulites, hypersthene amphibolites and gneisses. It forms in some hornfelses.

Grains from: (a) Grains are of pyroclastic origin from a beach sand, southern Turkey; (b) ophiolite-derived grains, beach sand, southern Turkey; (c) Upper Miocene, Hungary (Mmt 1.582).

Diopside–hedenbergite series

$CaMg[Si_2O_6]$ $CaFe^{2+}[Si_2O_6]$
monoclinic, biaxial (+)

$n\alpha$	1.664–1.695	$n\alpha$	1.716–1.726
$n\beta$	1.672–1.701	$n\beta$	1.723–1.730
$n\gamma$	1.694–1.721	$n\gamma$	1.741–1.751
δ	0.007–0.011	δ	0.018–0.020
Δ	3.22	Δ	3.56

Form in sediments: Grains are dominantly prismatic and appear either as stubby prisms or less frequently, long slender grains. Rounded edges and corners, also ragged terminations and other signs of dissolution are often exhibited. Cleavages and parting are usually well developed. Lamellar twinning or exsolution lamellae of orthopyroxene are seen occasionally. Some grains may contain finely distributed opaque inclusions.

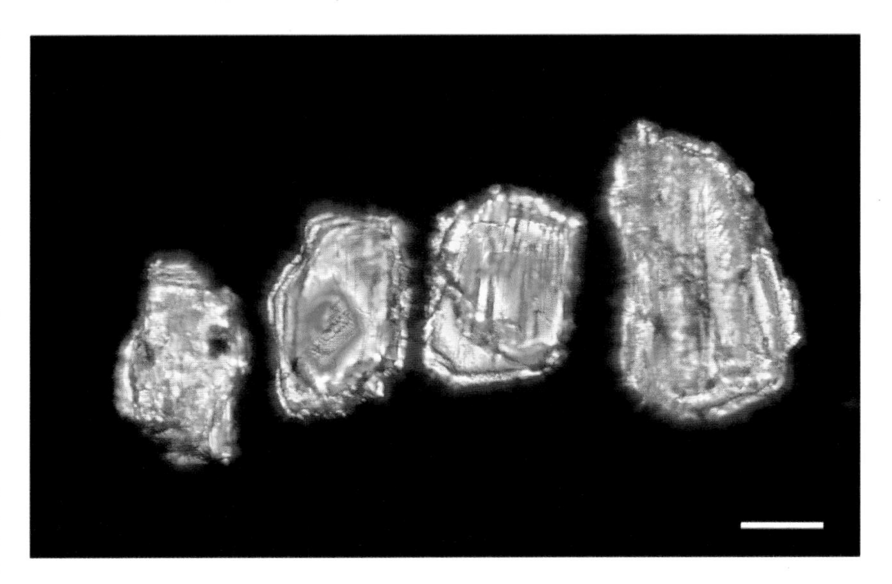

Colour: Colourless or pale green.

Pleochroism: Fe-rich members and hedenbergite may be pleochroic in shades of pale green, bluish green, pale greenish brown and yellow green.

Birefringence: Moderate to strong and interference colours are upper first- to second-order pink, yellow and bluish green. Some grains may exhibit bright interference colour bands.

Extinction: The extinction angle is large, ranging between 38° and 48°. {100} parting plates have parallel extinction.

Interference figure: Only poorly defined eccentric or flash figures can be obtained in detrital grains.

Elongation: Positive.

Distinguishing features: Prismatic morphology, cleavages, parting and large extinction angle define a clinopyroxene. As members of the diopside-hedenbergite series have similar optical properties to those of augites, they are often indistinguishable from each other by optical means. When only optical methods are used the colourless or very pale green detrital clinopyroxenes are usually identified as diopside and the highly coloured varieties as augite. However, it is more appropriate to denote them as diopsidic or augitic clinopyroxenes. Kyanite is optically negative and has a combination of cleavages and parting in right angles.

Occurrence: Clinopyroxenes of ultrabasic rocks are usually diopsidic in composition. Diopside and salite have been reported from alkali basalts, andesites, sub-alkaline magmas and from ultrabasic nodules in basic rocks. Hedenbergite occurs in some alkaline and acid igneous rocks, e.g. quartz-bearing syenites and occasionally granites. Diopside, salite and ferrosalite are formed by regional or contact metamorphism of Ca-rich sediments and are common in Ca- and Mg-rich schists of both igneous and sedimentary derivation. They are characteristic constituents of metasomatic rocks and skarns. Pyroxene granulites and gneisses may also contain diopside.

Grains from: Oligocene Sub-alpine Molasse, Beatenberg, Switzerland; Other grains: beach sand, southern Turkey (Mmt 1.582).

Diallage

monoclinic, biaxial (+)

The name diallage is applied to calcic clinopyroxenes (usually **diopside** or **augite**) which show a prominent and closely spaced parting on {100}.

Form in sediments: Owing to inherited large crystal size, diallage is more common in the coarser grades and it occurs as rather thin parting plates with a pearly lustre. It has a dominantly angular habit, as the thin plates are brittle and fracture easily, even upon mounting (lower). Thicker or massive grains may show slightly rounded corners. Some grains exhibit ragged edges or intensive etching. Exsolution lamellae are fairly common. Finely distributed opaque dust gives a speckled appearance to some grains. Parallel flake-like opaque inclusions on parting surfaces are called **Schiller-structure**. Serpentinized grains may also be encountered.

Colour: Colourless, pale green or pale yellow.

Pleochroism: Very weak or absent.

Birefringence: Moderate to strong and interference colours are pale greyish white or pale yellow in thinner grains, but range up to brilliant second-order orange, pink and blue in thick specimens.

Extinction: Of parting plates is parallel, but on rolling the grains in an immersion liquid onto (010), the large extinction angle, characteristic of clinopyroxenes, can be observed.

Interference figure: Parting plates yield a clear, off-centre optic axis figure with very large 2V, which is characteristic. Many bright isochromatic curves appear in thicker grains.

Elongation: Positive.

Distinguishing features: Platy habit, well-displayed {100} parting and parallel extinction, together with a characteristic optic axis figure are diagnostic of diallage. Grains with poorly defined parting may be mistaken for enstatite, but the dominantly stout prismatic morphology of the former, and the characteristic optic axis figure of the latter, aid distinction. Some thin prismatic anthophyllitic amphibole may resemble diallage, but the relief of the former is lower and it lacks parting. Platy celestite has weaker birefringence and parting is absent.

Occurrence: Diallage is a characteristic mineral of coarse-grained gabbros, diallagites and some other ultrabasic rocks.

Remarks: Diallage is probably more common in sediments than was previously reported. Owing to its platy habit, pearly lustre and sometimes turbid appearance, it can be mistaken for some members of the sulphate group. Diallage may be anticipated in the presence of other pyroxenes, hornblende and serpentine, generally when the mineral suite indicates basic or ultrabasic source rocks.

Grains from: Upper, top left: Beach sand, southern Turkey; Other grains: Oligocene Sub-alpine Molasse, Beatenberg, Switzerland (Mmt 1.582).
Lower: Beach sand, southern Turkey (CB).

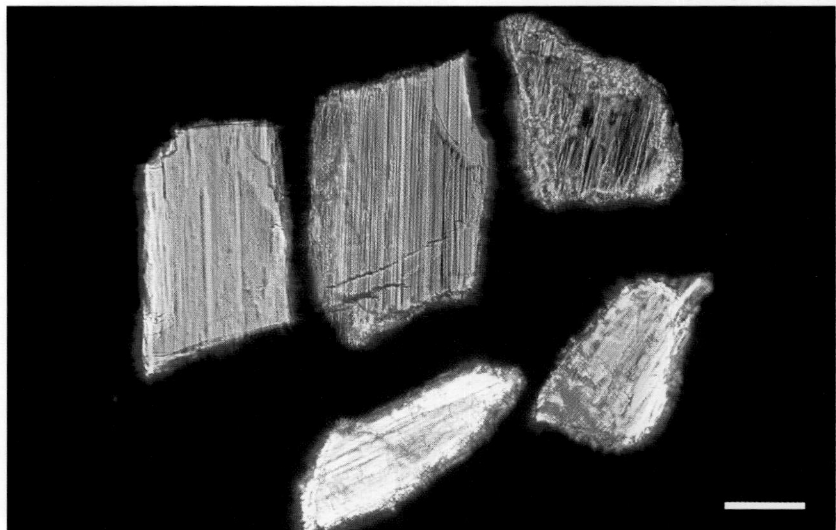

Augite

$(Ca,Mg,Fe^{2+},Fe^{3+},Ti,Al)_2[(Si,Al)_2O_6]$
monoclinic, biaxial (+)

$n\alpha$	1.671–1.735
$n\beta$	1.672–1.741
$n\gamma$	1.703–1.774
δ	0.032–0.039
Δ	2.96–3.52

Form in sediments: Augite is the commonest detrital pyroxene. In volcanic rocks it is dominantly euhedral to subhedral, and grains eroded from them often retain their form, occurring as short or long slender prisms with terminations at one or both ends (a, b), and sometimes square to octagonal basal sections. Volcanic augites may show embayments or corrosion. The most common detrital morphologies are stubby prismatic fragments, rounded forms and irregularly shaped particles. These may show rectangular breakage patterns, sometimes with conchoidal fractures or grooves on their surface. Compositional zoning is fairly common. Twinning is either simple or repeated. Augites of plutonic rocks often enclose exsolution lamellae of ortho- and other clinopyroxene phases. Accessory mineral and fluid inclusions are frequent. Cleavages in detrital augites are not prominent. Owing to the unstable nature of augite in sediments, post-depositional dissolution frequently produces etching of the grains which is manifested in 'hacksaw' terminations (e). Advanced stages of etching produce skeletal forms (f) before complete dissolution.

Colour: Augite appears in various shades of green, and sometimes brown or yellowish brown. Titanaugite is highly coloured in purple brown or violet brown (b). Colour zoning, frequently seen in volcanic varieties, is manifested by paler cores and darker outer zones (a).

Pleochroism: Mg-rich augites are non-pleochroic, but increasing iron content results in deeper colours and pleochroism in pale blue, green and yellow green. Titanaugite is distinctly pleochroic: α, pale violet brown; β, violet pink; γ, bluish violet.

Birefringence: Moderate but increases with higher iron content. {100} partings have low birefringence. The second- and third-order interference colours appear as bright yellow, orange, red and blue concentric bands on rounded grains, and parallel with the outline on prismatic specimens. Highly coloured varieties show abnormal interference colours because of the masking effect of natural mineral colour.

Extinction: The γ : z maximum value, measured on {010} faces, ranges between 35°C and 48°C. **N.B.**: fragments laying parallel to the {100} parting have straight extinction. Owing to compositional zoning some fragments may show mottled extinction.

Interference figure: Augite grains rarely yield appreciable interference figures. Cleavage fragments that lie on {110} provide off-centre figures, usually with several isochromatic curves. {100} partings show off-centre optic axis figures.

Elongation: Positive.

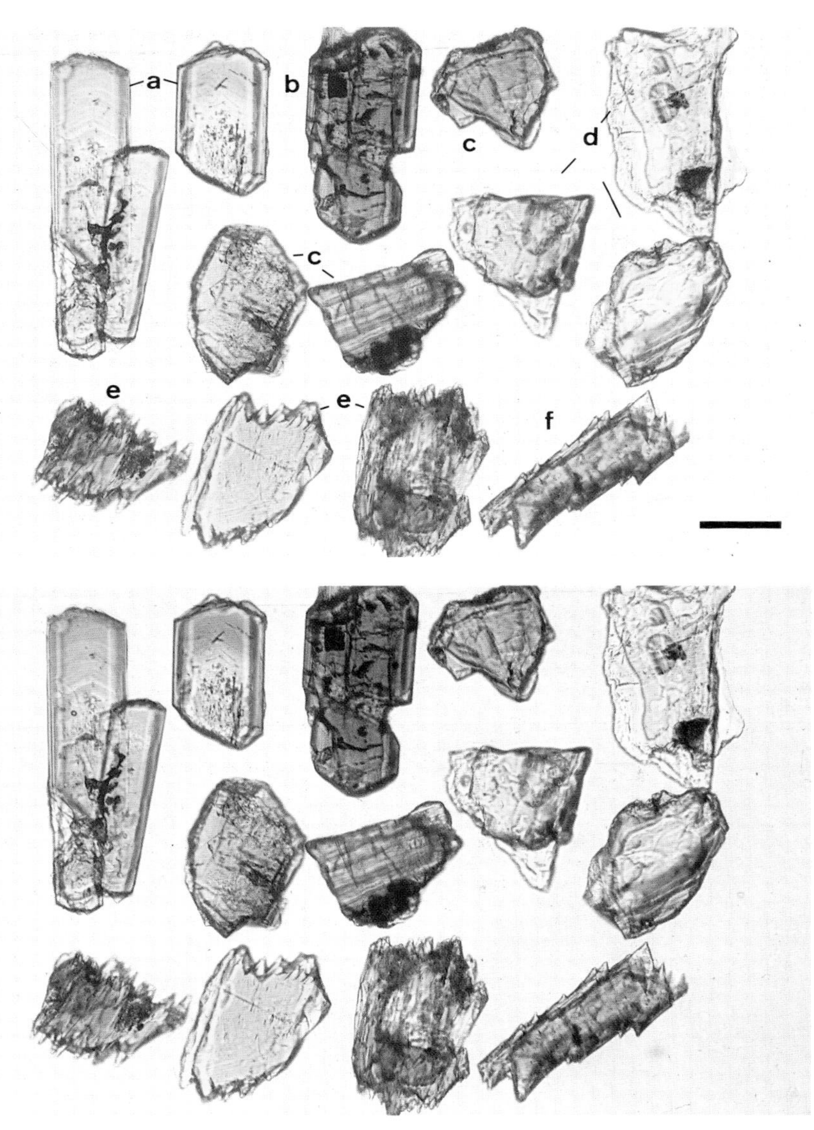

Distinguishing features: Fairly high relief, large extinction angle and green, sometimes brownish, colours are characteristic of augite. Owing to compositional variations, only chemical analysis can provide an accurate determination of the augite variety and also a positive distinction of augite from diopside. The relief of augite is higher than that of diopside and the $\gamma : z$ extinction angle of the latter is smaller. When only optical diagnosis is made, the name augite is generally used for green or brownish clinopyroxenes, whereas the pale green or colourless clinopyroxenes are designated as diopside. However, it is more appropriate to name these species as augitic or diopsidic clinopyroxenes. Titanaugite can be distinguished by its characteristic colour and pleochroism. Aegirine-augite is highly coloured, pleochroic and frequently shows negative elongation. Orthopyroxenes give parallel extinction in all longitudinal sections. Amphiboles have a lower relief, smaller extinction angle and exhibit better cleavages than detrital clinopyroxenes.

Occurrence: Augite is widespread in various ultramafic and intermediate igneous rock types and is particularly common in gabbros, dolerites, andesites and basalts, and also in some peridotites. It is less frequent in metamorphic rocks, where it is found in dark-coloured gneisses and pyroxene granulites. Titanaugite is typical in basic alkaline rocks.

Grains from: (a) and (e) Tertiary infilling of bauxite pockets, Seydisehir, Turkey; (b) beach sand, Tenerife, Canary Islands; (c) beach sand, southern Turkey; (d) beach sand, Luzon, Philippines; (f) Oligocene, Barrême Basin, France (Mmt 1.582).

Aegirine-augite

$(Na,Ca)(Fe^{3+},Fe^{2+},Mg,Al)[Si_2O_6]$
monoclinic, biaxial (−) (+)

$n\alpha$	1.700–1.750
$n\beta$	1.710–1.780
$n\gamma$	1.730–1.800
δ	0.030–0.050
Δ	3.40–3.55

Form in sediments: Generally aegirine-augite occurs as short, stubby prisms, angular fragments or occasionally slightly rounded grains. Pyroxene cleavages are frequently exhibited. Grains may contain minute accessory mineral inclusions or opaque impurities.

Colour: Green, yellowish or brownish green, usually with a slight bluish tinge. Colour zoning, manifested in darker internal portions and paler periphery is common.

Pleochroism: Distinct: α, bright green, bluish green to olive green; β, yellow green, grass green; γ, pale green, pale brownish green.

Birefringence: Rather strong. Interference colours range from upper first-order to second-order shades and are usually masked by strong inherent mineral colour.

Extinction: The extinction angle varies from 0° to 21°, but is symmetrical in cross-sections.

Interference figure: Cleavage fragments yield poor interference figures. Usually a flash of one diffuse isogyre appears in a dark-green field. Basal sections give off-centre bisectrix figures with large 2V. r>v dispersion is strong.

Elongation: Negative.

Distinguishing features: Deep colour, pleochroism, small extinction angle and length-fast character are diagnostic and assist in distinguishing aegirine-augite from other clinopyroxenes. Aegirine grains are generally long slender prisms, but optical properties alone are insufficient to distinguish between aegirine and aegirine-augite. Aegirine-augite can be distinguished from amphiboles by their length-slow character and better cleavages. It may resemble some alkali amphiboles, but they have a markedly different pleochroic scheme. Tourmaline is uniaxial, has a lower relief, stronger pleochroism and poor cleavages. The colour of epidote is a characteristic yellowish or pistachio green, its morphology is highly irregular and it displays higher-order polarization colours.

Occurrence: Aegirine-augite occurs in soda-rich igneous rocks such as syenites and various ultra-alkaline types. Na-rich regionally metamorphosed rocks may contain aegirine-augite resulting from Na-metasomatism. Sodic pyroxenes are usually found together with sodic amphiboles.

Grains from: Oligocene Molasse, borehole Leymen-1, 252 m, France (Mmt 1.582).

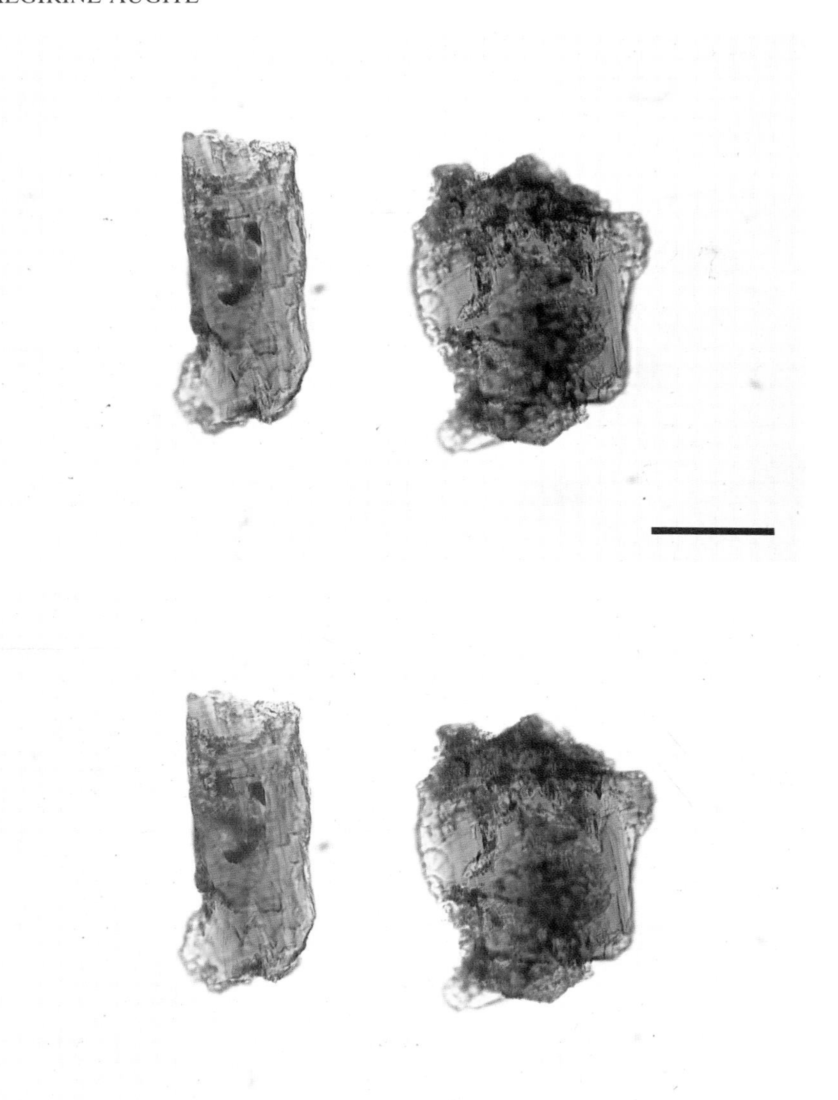

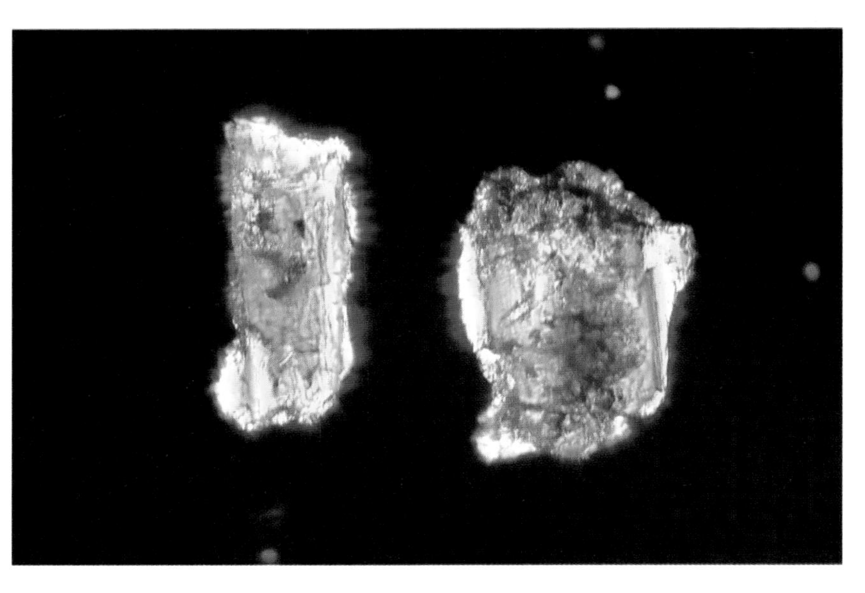

Aegirine (acmite)

$Na,Fe^{3+}[Si_2O_6]$
monoclinic, biaxial (−)

$n\alpha$	1.750–1.776
$n\beta$	1.780–1.820
$n\gamma$	1.800–1.836
δ	0.040–0.060
Δ	3.55–3.60

Form in sediments: Aegirine (acmite) occurs as euhedral crystals, thin long prismatic fragments or rarely as grains composed of radiating fibres. Worn grains may also occur. Fluid, opaque and accessory mineral inclusions are often present.

Colour: Bright green, usually with a bluish tinge. Grains frequently display colour zoning or irregular colour distribution. **Acmite** is distinguished from aegirine on the basis of its brown or brownish-green colour.

Pleochroism: is distinctive: α, emerald green, deep green; β, grass green, deep green; γ, brownish green, yellowish brown. Acmite is weakly pleochroic in pale brown and yellow.

Birefringence: Maximum birefringence is strong, and interference colours are third- or fourth-order green, yellow and dark pink, but they are often masked by the strong mineral colour.

Extinction: The extinction angle is small, ranging from 2° to 10°.

Interference figure: {100} faces show off-centre bisectrix figures and {010} faces yield flash figures. Isogyres appear somewhat obscured on a yellow-green field. Colour bands are indistinct.

Elongation: Negative.

Distinguishing features: Dominantly euhedral or slender prismatic morphology, vivid colour, distinct pleochroism, combined with optically negative character and negative elongation, are diagnostic. These properties serve to distinguish it from other pyroxenes. Aegirine-augite has a larger extinction angle and is optically either positive or negative. However, in addition to optical means it is advisable to use chemical or X-ray methods to ensure a positive diagnosis of aegirine. The length-fast character distinguishes aegirine from amphiboles. Slender green tourmaline has a weaker relief and lower-order interference colours. Prismatic fragments of epidote may resemble aegirine, but the colour and pleochroic scheme of epidote is always in a yellowish-green shade in contrast with the dominant bluish or brownish-green colours of aegirine.

Occurrence: Aegirine crystallizes in alkali-rich intrusives, such as syenite, nepheline syenite, syenite pegmatite and alkali-rich granite. Na-rich regionally metamorphosed rocks occasionally contain aegirine. It is uncommon, but may appear in alkali extrusive rocks. Aegirine and acmite have been reported as authigenic (Milton & Eugster 1959, Fortey & Michie 1978).

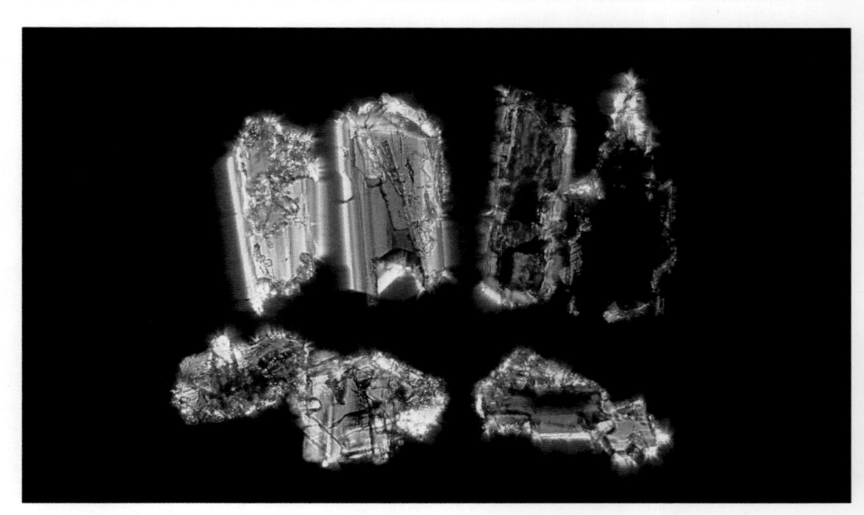

Grains from: Pyroclastic origin from the Palaeogene Thanet Formation, offshore south-east England, borehole 79/7A 57.81 m. Grain top right is intergrown with arfvedsonite (Mmt 1.882). Courtesy of A. C. Morton.

AMPHIBOLE GROUP

General considerations

Amphiboles are common rock-forming minerals of many igneous and metamorphic rocks and are widespread in detrital sediments. They constitue an extremely complex group in which a considerable variation of chemistry and physical properties is characteristic. There is a strong relationship between these properties and the conditions of crystallization.

The chemical complexity of amphiboles implies that optical observation alone is inadequate to diagnose a particular species positively and, for a precise identification, chemical analyses are essential. These can be performed easily on detrital specimens, as concentrating amphiboles by magnetic separation or by handpicking for further study is relatively simple. Electron-probe microanalysis will then provide the data for calculation. The nomenclature based on crystal chemistry, and introduced by the Report of the IMA Subcommittee on amphiboles (Leake 1978, discussed also by Hawthorne 1981) facilitiates a rapid allocation of amphibole analyses. The nomenclature uses well-established names which have been chemically codified. Specific elemental ranges are indicated by prefixes and adjectival modifiers.

For amphiboles diagnosed only by their optical properties, the Report recommends that the assigned amphibole name be made into an adjective, followed by the word amphibole, e.g. tremolitic amphibole, glaucophanitic amphibole, etc. In the case of detrital amphiboles it is often necessary to adopt this approach. In most instances, allocation of a particular sub-group can be made by optical observation. Parallel extinction indicates orthorhombic amphiboles; intensive colours in bluish shades, strong pleochroism and often negative elongation are characteristic of alkali amphiboles; tremolitic amphiboles lack colour; and actinolitic amphiboles are generally fibrous. When appropriate, for calcic amphiboles the well-established name 'hornblende' continues to be used, but with an adjective. This series is typified by intensive green, bluish-green and brown colours (hence the often used names blue-green, green-brown, and brown or oxyhornblende).

Amphibole-rich heavy mineral assemblages, in which the amphibole species have been precisely determined, can prove highly valuable in specifying source-rock lithologies and distinguishing or correlating sand bodies.

Because of their low chemical stability, amphiboles are easily affected and eliminated by weathering and diagenetic processes. Deposits older than Tertiary are often devoid of amphiboles, though there are reports of well-preserved amphiboles in many ancient successions.

Detrital amphiboles, used for the present illustration, were derived from diverse parageneses and have been selected from a wide variety of ancient and modern sedimentary environments. The purpose was to highlight the most commonly occurring species in sediments and, whenever possible, show several representatives of a particular group. With a few exceptions, all grains have been identified and named by their chemical content.

In the following classification and description of detrital amphiboles the principle of the nomenclature discussed above has been adopted. For simplicity, only the chemical formulae of the end members are indicated. Further details can be found in the Report (Leake 1978).

The standard amphibole formula is:

$$A_{0-1}B_2C^{vi}{}_5T^{iv}{}_8O_{22}(OH,F,Cl)_2$$

where $A = Na,K$; $B = Na,Li,Ca,Mn,Fe^{2+},Mg$;
$C = Mg,Fe^{2+},Mn,Al,Fe^{3+},Ti$;
$T = Si,Al$

The four principal amphibole groups on the basis of the numbers of atoms of $(Ca + Na)_B$ and Na_B are defined as:

1 iron–magnesium–manganese amphiboles
2 calcic amphiboles
3 sodic–calcic amphiboles
4 alkali amphiboles

1 IRON–MAGNESIUM–MANGANESE AMPHIBOLES

Orthorhombic forms

End members:

Magnesio-anthophyllite	$Mg_7Si_8O_{22}(OH)_2$
Ferro-anthophyllite	$Fe^2Si_8O_{22}(OH)_2$
Sodium-anthophyllite	$Na(Mg,Fe^2)_7AlSi_7O_{22}(OH)_2$
Magnesio-gedrite	$Mg_5Al_2Si_6Al_2O_{22}(OH)_2$
Ferro-gedrite	$Fe^2{}_5Al_2Si_6Al_2O_{22}(OH)_2$
Sodium-gedrite	$Na(Mg,Fe^2)_6AlSi_6Al_2O_{22}(OH)_2$
Magnesio-holmquistite*	$Li_2Mg_3Al_2Si_8O_{22}(OH)_2$
Ferro-holmquistite*	$Li_2Fe_3Al_2Si_8O_{22}(OH)_2$

* Not discussed in the present work.

Monoclinic forms Monoclinic forms include the **cummingtonite–grunerite series**, beside other members. Cummingtonite may occur locally in sediments shed by cummingtonite-bearing amphibolites and some metabasic rocks. Grunerite forms largely in metamorphosed iron formations. The characteristic feature of members of this series is the multiple twinning on {100} with narrow twin lamellae. Cummingtonite is optically biaxial positive and grunerite is biaxial negative.

Anthophyllite and gedrite

orthorhombic, biaxial (−) (+)

$n\alpha$	1.596–1.694
$n\beta$	1.605–1.710
$n\gamma$	1.615–1.722
δ	0.013–0.028
Δ	2.85–3.57

Form in sediments: Grains are either prismatic, bladed or composed of variously oriented fibres or fine needles. The prismatic grains often exhibit cross fractures. Inclusions are mainly opaque impurities.

Colour: Mg-anthophyllites are colourless, but with increasing iron content grains may be light brown or light yellowish brown. Gedrite has stronger colours in pale brown and clove brown (lower right).

Pleochroism: Coloured varieties display weak to moderate pleochroism in: α, pale grey brown; β, brownish grey to clove brown; γ, dark brown, grey, greyish green.

Birefringence: Maximum birefringence is weak to moderate and interference colours are vivid second-order tints. Prismatic grains often display many bright interference colour bands parallel with the long axis of the crystal.

Extinction: Of prisms is parallel and that of fibrous grains is undulatory.

Interference figure: Cleavage fragments provide flash-figures with widely spaced isogyres. {100} faces yield clear acute bisectrix figures. Mg-anthophyllites are optically negative, but those with higher iron content are optically positive. Gedrite has an optically positive character, but grains seldom yield an interference figure.

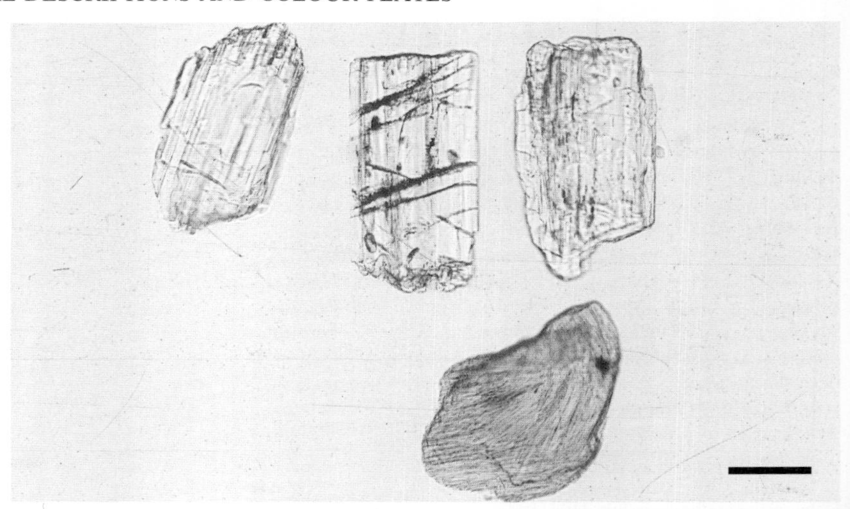

Elongation: Positive.

Distinguishing features: Prismatic anthophyllite is characterized by very pale (or lack of) colours, amphibole cleavages and parallel extinction. Tremolite strongly resembles anthophyllite, but it has oblique extinction. The RI of enstatite is higher and it often exhibits exsolution lamellae, serpentinization or typical pyroxene etch-features called 'hacksaw terminations'. Diallage displays prominent parting and slightly inclined extinction. Prismatic sillimanite can be confused with anthophyllite, but the former lacks amphibole cleavages. Fibrous colourless anthophyllite cannot be distinguished from the fibrous sillimanite variety, fibrolite, by optical means alone. Pale-coloured anthophyllites and brown gedrites are, however, easily distinguishable from the always colourless fibrolite.

Occurrence: Anthophyllite and gedrite are typically metamorphic minerals and they appear in amphibolites, gneisses and granulites, as well as in regionally metamorphosed ultrabasic rocks. Mg- and Fe-metasomatism may form anthophyllite or gedrite in argillaceous sediments.

Grains from: Upper row: Anthophyllitic amphibole, Middle America Trench, DSDP Leg-66; Lower right: Gedrite, river sand, Danube, Hungary (Mmt 1.582).

2 CALCIC AMPHIBOLES

Orthorhombic forms

End members:

Tremolite	$Ca_2Mg_5Si_8O_{22}(OH)_2$
Ferro-actinolite	$Ca_2Fe^2{}_5Si_8O_{22}(OH)_2$
Edenite*	$NaCa_2Mg_5Si_7AlO_{22}(OH)_2$
Ferro-edenite	$NaCa_2Fe^2{}_5Si_7AlO_{22}(OH)_2$
Pargasite	$NaCa_2Mg_4AlSi_6Al_2O_{22}(OH)_2$
Ferro-pargasite	$NaCa_2Fe^2{}_4AlSi_6Al_2O_{22}(OH)_2$
Hastingsite	$NaCa_2Fe^2{}_4Fe^3Si_6Al_2O_{22}(OH)_2$
Magnesio-hastingsite	$NaCa_2Mg_4Fe^3Si_6Al_2O_{22}(OH)_2$
Alumino-tschermakite	$Ca_2Mg_3Al_2Si_6Al_2O_{22}(OH)_2$
Ferro-alumino-tschermakite	$Ca_2Fe^2{}_3Al_2Si_6Al_2O_{22}(OH)_2$
Ferri-tschermakite	$Ca_2Mg_3Fe^3{}_2Si_6Al_2O_{22}(OH)_2$
Ferro-ferri-tschermakite	$Ca_2Fe^2{}_3Fe^3{}_2Si_6Al_2O_{22}(OH)_2$
Alumino-magnesio-hornblende	$Ca_2Mg_4AlSi_7AlO_{22}(OH)_2$
Alumino-ferro-hornblende	$Ca_2Fe^2{}_4AlSi_7AlO_{22}(OH)_2$
Kaersutite	$NaCa_2Mg_4TiSi_6Al_2(O+OH)_{24}$
Ferro-kaersutite	$NaCa_2Fe^2{}_4TiSi_6Al_2(O+OH)_{24}$

* Members between edenite and kaersutite are described within the 'hornblende series'.

3 SODIC–CALCIC AMPHIBOLES

The sodic–calcic amphibole group includes a series of monoclinic amphiboles containing 18 end-members. The chemically defined names **winchite**, **barroisite**, **richterite**, **katophorite** and **taramite** are specified by appropriate prefixes to indicate elemental ranges.

As detrital grains sodic–calcic amphiboles may have local importance, but their identity can only be revealed by chemical means. In a study of sodic and calcic amphibole-rich assemblages of the Lower Miocene western Alpine Molasse Mange-Rajetzky & Oberhänsli (1982, Figure 7) analysed grains with chemical compositions plotting in the **richterite** and **ferro-winchite** fields respectively. Morton (1983) reported **magnesio-katophorite** of volcaniclastic origin from the Palaeogene Thanet Formation, England.

Tremolite

monoclinic, biaxial (−)

$n\alpha$	1.599–1.688
$n\beta$	1.612–1.697
$n\gamma$	1.622–1.705
δ	0.027–0.017
Δ	3.02–3.44

Form in sediments: Grains most commonly appear as thin, short stumpy or long slender prisms, and their ends often show etch features. Some grains may enclose opaque impurities.

Colour: Colourless, pale greyish or greenish white.

Pleochroism: Non-pleochroic.

Birefringence: Weak to moderate, and interference tints are first-order grey, yellow or orange. Thicker grains may exhibit second-order polarization colours.

Extinction: On {010} sections the extinction angle varies between 15° (Fe-tremolite) and 21° (Mg-tremolite).

Interference figure: {100} crystal faces yield a clear acute bisectrix figure with large 2V. Colour bands are usually absent.

Elongation: Positive.

Distinguishing features: Prismatic habit together with amphibole cleavages, lack of colour, frequent ragged edges and low-order interference tints are diagnostic. Tremolite may be confused with anthophyllite, but the extinction of the latter is parallel and it displays more intensive polarization colours. Diopside has higher RI and a large extinction angle. Clinozoisite appears with a higher relief and shows vivid, often anomalous, interference colours. The extinction of sillimanite is parallel, its birefringence is stronger and it lacks amphibole cleavages.

Occurrence: Tremolite is most commonly a product of thermal or regional metamorphism of siliceous dolomites. Some tremolite-bearing assemblages are associated with low-grade regionally metamorphosed ultrabasics, such as tremolite–talc–antigorite and tremolite–talc schists.

Grains from: Oligocene, Barrême Basin, France (Mmt 1.539).

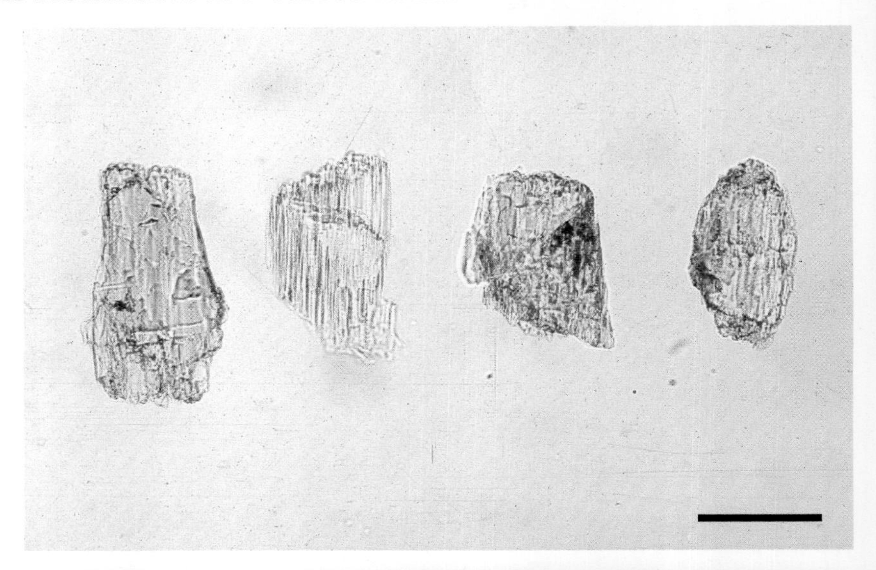

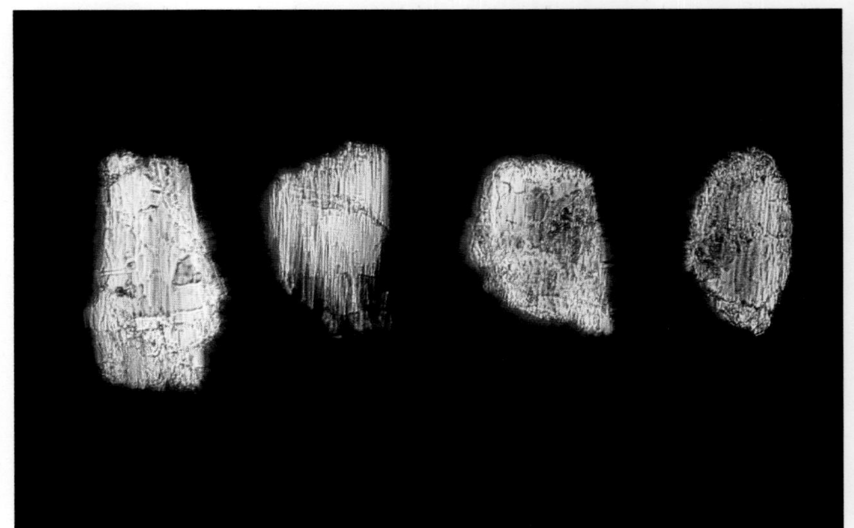

Ferroactinolite, actinolite

monoclinic, biaxial (−)

$n\alpha$	1.599–1.688
$n\beta$	1.612–1.697
$n\gamma$	1.622–1.705
δ	0.027–0.017
Δ	3.02–3.44

Form in sediments: Actinolite occurs as fibrous grains composed of parallel, subparallel or radially arranged fibres, or sometimes interwoven needles. Columnar or bladed crystals are less common. Rounding of the grains may be discernible. Iron ore and carbonaceous inclusions are occasionally present.

Colour: Pale green, pale yellowish green to bluish green.

Pleochroism: Certain grains show distinct pleochroism: α, pale yellow, pale brown, yellowish green; β, pale yellowish green or green; γ, deep greenish blue, pale green; others are non-pleochroic.

Birefringence: Weak to moderate and is influenced by the chemistry of the mineral. Interference colours are second-order yellow, deep orange, blue and bluish green.

Extinction: On columnar and bladed crystals the extinction angle is between 11° and 15°.

Interference figure: Because of the dominantly fibrous habit of the mineral, it is difficult to obtain an interference figure.

Elongation: Positive.

Distinguishing features: The fibrous habit and greenish colour serve to distinguish actinolite from other amphiboles, especially from the green hornblende varieties and gedrite, and from colourless tremolite. Iron-rich actinolites, merging to hornblende, necessitate chemical tests for a positive identification. Some etched, pseudo-fibrous blue-green hornblende may resemble actinolite, but its extinction angle is larger and the etch patterns characteristically develop in the direction of prominent cleavages.

Occurrence: Actinolite is generated by low-grade regional metamorphism (greenschist facies). It forms in schists derived from basic rocks and also in high-pressure parageneses together with blue sodic amphiboles. In some impure lime-silicate rocks actinolite is produced by regional or contact metamorphism. Fibrous **uralite** is actinolitic in composition and is formed by hydrothermal alteration of pyroxenes.

Grains from: Actinolitic amphibole, beach sand, St Ives, Cornwall, England (Mmt 1.582).

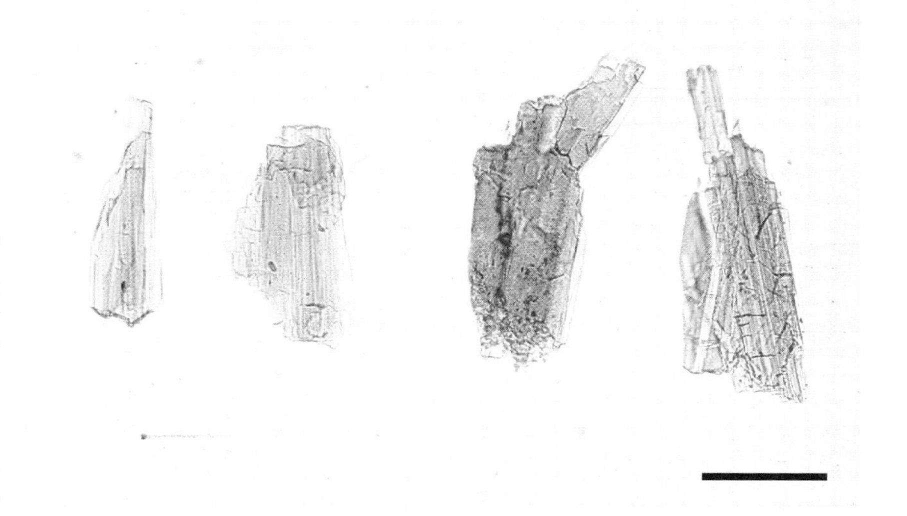

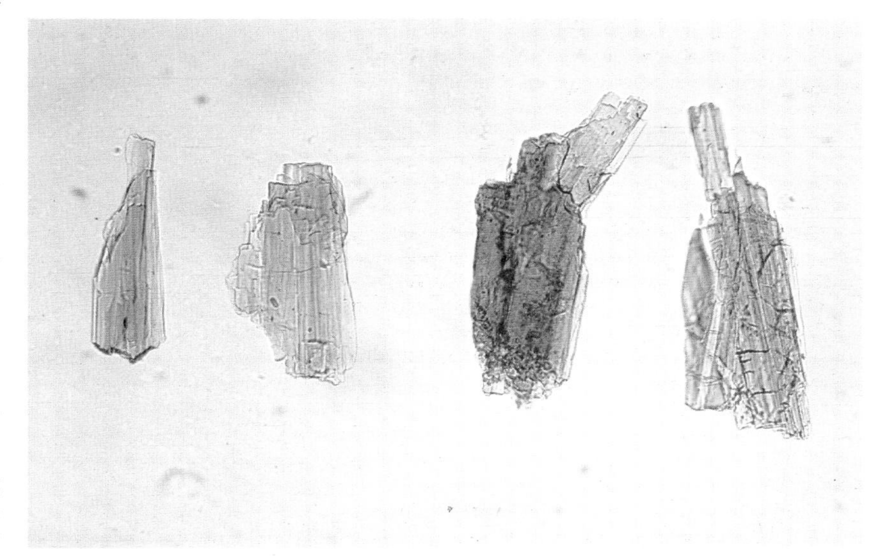

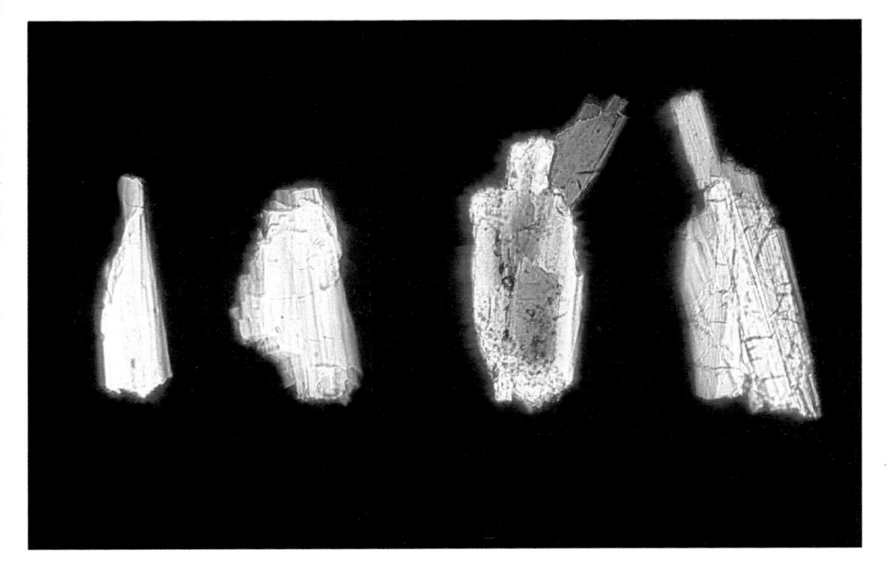

Hornblende series

monoclinic

Edenite and pargasite are optically positive; other species are optically negative.

	edenite ferro-edenite tschermakite ferro-tschermakite	pargasite ferro-pargasite
$n\alpha$	1.615–1.705	1.613
$n\beta$	1.618–1.714	1.618
$n\gamma$	1.632–1.730	1.635
δ	0.014–0.026	0.022
Δ	3.02–3.45	3.05

	hastingsite	oxyhornblende (magnesio- or magnesian hastingsite)	kaersutite
$n\alpha$	1.702	1.662–1.690	1.670–1.689
$n\beta$	1.729	1.672–1.730	1.672–1.730
$n\gamma$	1.730	1.680–1.760	1.680–1.760
δ	0.028	0.018–0.070	0.018–0.083
Δ	3.50	3.19–3.30	3.20–3.28

Members of the hornblende series are the most widespread detrital amphiboles. When chemical analysis is not available, sedimentary petrologists tend to distinguish hornblende varieties according to their colour rather than any other properties. Thus, such names as green-brown, blue-green and brown (basaltic, or oxyhornblende) commonly appear in reports of detrital mineral analyses.

Form in sediments: Grains are predominantly cleavage fragments determined by perfect {110} prismatic cleavage. This also governs grain orientation upon mounting. The morphology varies from short or slender prisms, irregular or rectangular fragments, to long thin flakes. Some grains may be thick and massive, platy or bladed; others are partially fibrous or are sometimes intergrown with another amphibole, rarely with pyroxene phases. The ends of the prisms are often rounded, abraded but most commonly, especially in older sediments, they show signs of various degrees of dissolution. This is recognized by ragged edges, various etch features or skeletal forms. The directions of cleavages and cross fractures are usually well displayed. Striated crystal faces may also appear. Grains of volcanic origin are often euhedral and have terminations at one or both ends. Those of aeolian deposits are well rounded. Inclusions of accessory minerals, as well as epidote, biotite, opaque particles and fluid globules, are common.

Colour: Characteristic hornblende colours are bluish green, brown green and brown. The dark-brown or reddish-brown varieties are less common. Colour zoning or patchy colour arrangements are frequent. Very dark-brown grains may appear opaque. Decomposition of iron-rich varieties often results in a partial iron-oxide coating.

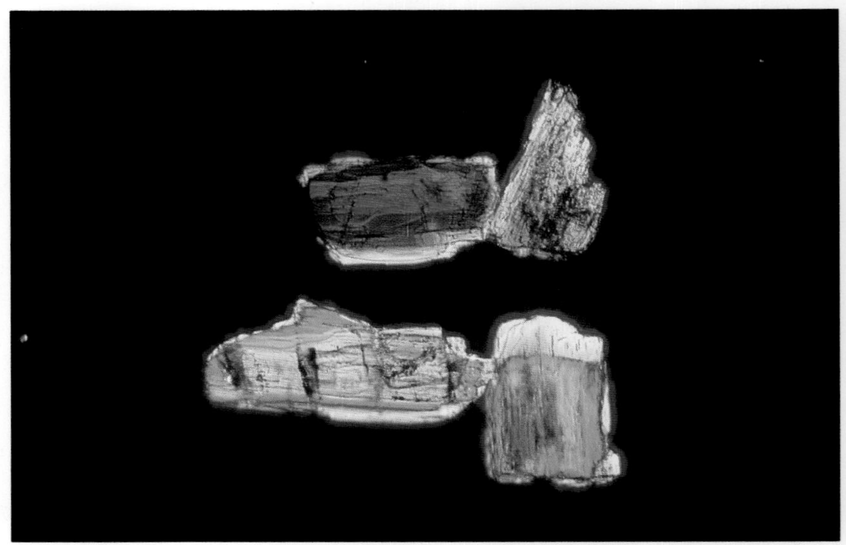

Pleochroism: Strong. The strongest absorption is in the β and γ vibration direction and, as grains are dominantly {110} cleavage fragments, α is rarely discernible. The most common pleochroic scheme of the *green coloured varieties* is as follows:

α colourless, yellowish green, pale yellow, yellow green;
pale yellow, pale brown green, light green, green;
pale green, bluish green.

Brown varieties usually show strong pleochroism in shades of:

α yellow green, pale yellow, greenish yellow, greenish brown;
β pale brown, yellowish brown, reddish brown;
γ brown, deep reddish brown, red brown.

Birefringence: Moderate to strong. Polarization colours are variable and largely depend on thickness and mineral colour. Shades are either upper first-order, second- and lower third-order tints, and are often obscured by strong inherent mineral colour.

Extinction: The γ : z extinction angle measured on {010} ranges between 12° and 34°. Occasional {100} parting plates have parallel extinction. The extinction angle of oxyhornblende and kaersutite is 0°–18°.

Interference figure: Grains generally yield poor, highly eccentric bisectrix figures. Sharp isogyres appear in a rather dark field with a few isochromes. Some {100} sections of the brown varieties and also actinolitic hornblende display well-centred or slightly off-centre acute bisectrix figures.

Elongation: Positive.

Distinguishing features: The characteristic prismatic morphology, amphibole cleavages and strong colours provide an easy diagnosis of the hornblende series, but distinction of a particular member within the group requires chemical evidence. Intense colours and higher RI distinguish hornblende from iron-free amphiboles (e.g. tremolitic and anthophyllitic). Sodic amphiboles are usually blue, and several varieties are length fast. Confusion may arise from the resemblance of some hornblende grains to tourmaline, calcic pyroxenes or aegirine. The deep brown varieties may be mistaken for allanite. Tourmaline is uniaxial, length fast and has poor cleavages. Its greater absorption is normal to the vibration plane of the lower nicol. Calcic pyroxenes have stronger relief, a larger extinction angle and less distinct cleavages. Aegirine has higher refractive indices, poorer cleavages and is length fast. The morphology of allanite is usually irregular; it lacks good cleavages and has a markedly higher relief.

Occurrence: Members of the hornblende series form in a wide range of P,T conditions; hence they are present in a large variety of igneous and metamorphic rocks. Of these, the common source rocks are mentioned here. In igneous rocks, hornblendes are widespread in intermediate, acid and alkaline intrusives (e.g. diorites, granodiorites, tonalites, nepheline syenite, essexite). In ultramafic rocks, the edenitic and tschermakitic compositions are the principal occurrences, whereas in pegmatites and alkali plutonics hastingsite forms. Of volcanics, basalts, trachites, andesites, latites, basanites, teffrites and their tuffs contain oxyhornblende varieties. The titanium-rich kaersutite is generated in alkaline volcanics

(trachybasalts, trachy-andesites, trachytes and alkali rhyolites). In regionally metamorphosed rocks, ranging from greenschist facies to the lower part of the granulite facies, and in metasomatic rocks, hornblende species are common constituents. They are the principal rock-forming minerals of the amphibolite facies. The most iron-rich members (e.g. hastingsite, ferro-edenite, ferro-pargasite) form in schists and gneisses.

Grains from: First page, upper, top row: Edenitic hornblende, glacial sand, Maloya, Switzerland; Bottom row: Magnesio-hornblende, grab sample, Port of Beira, Mozambique.
First page, lower: Ferro-edenitic hornblende, river sand, Danube, Hungary.
Second page, upper: Magnesio-hastingsitic hornblende, Oligocene, Barrême Basin, France.
Second page, lower: Hastingsite, Tertiary infilling of bauxite pockets, Seydisehir, Turkey.
This page, two grains left: Magnesian hastingsitic hornblende, Middle America Trench, DSDP Leg 66; Other grains: kaersutite, beach sand, Tenerife, Canary Islands.
All grains are embedded in Mmt 1.582.

4 ALKALI AMPHIBOLES

End members:	
Glaucophane	$Na_2Mg_3Al_2Si_8O_{22}(OH)_2$
Ferro-glaucophane	$Na_2Fe^{2}{}_3Al_2Si_8O_{22}(OH)_2$
Magnesio-riebeckite	$Na_2Mg_3Fe^{3}{}_2Si_8O_{22}(OH)_2$
Riebeckite	$Na_2Fe^{2}{}_3Fe^{3}{}_2Si_8O_{22}(OH)_2$
Eckermannite	$NaNa_2Mg_4AlSi_8O_{22}(OH)_2$
Ferro-eckermannite	$NaNa_2Fe^{2}{}_4AlSi_8O_{22}(OH)_2$
Magnesio-arfvedsonite	$NaNa_2Mg_4Fe^{3}Si_8O_{22}(OH)_2$
Arfvedsonite	$NaNa_2Fe^{2}{}_4Fe^{3}Si_8O_{22}(OH)_2$
Kozulite	$NaNa_2Mn_4(Fe^{3},Al)Si_8O_{22}(OH)_2$

Glaucophane–riebeckite series

Members: Glaucophane, ferro-glaucophane, crossite, magnesio-riebeckite, riebeckite.

Al–Fe^{3+} substitution characterizes members of this series from glaucophane through crossite to magnesio-riebeckite and there is also a complete solid solution between magnesio-riebeckite and riebeckite.

	glaucophane monoclinic biaxial (−)	crossite monoclinic biaxial (−)	riebeckite monoclinic biaxial (−) (+)
$n\alpha$	1.594–1.647	1.647–1.690	1.690–1.702
$n\beta$	1.612–1.663	1.663–1.690	1.690–1.712
$n\gamma$	1.618–1.663	1.663–1.702	1.702–1.719
δ	0.008–0.022	0.012–0.016	0.006–0.016
Δ	0.08–3.30	3.1–3.3	3.02–3.42

Form in sediments: Grains are dominantly {110} cleavage flakes and they appear as slender or stumpy prisms elongated in the direction of the *c* axis, or less frequently rectangular fragments. Fibrous columnar grains, mats of fibres and acicular aggregates also occur. Small crystals can be found as constituents of composite grains formed of clinozoisite, albite, mica, etc. Cleavage traces and cross fractures are usually visible. Ragged edges are not as pronounced as on calcic amphiboles.

Colour: Glaucophane close to its end-member composition has light pale lavender colours owing to a low Fe-content. Iron-rich glaucophane and other members of the series display vivid colours in shades of blue. Riebeckite is bright bluish green with an emerald-green hue. Resulting from frequent compositional zoning, colours are often arranged in concentric layers or there are darker cores with paler rims, and vice versa. Patchy colour patterns are also common.

Pleochroism: Intense pleochroism is characteristic. The generalized pleochroic scheme is:

α light yellow, colourless, pale mauve;
β lavender blue, deep blue, purplish blue;
γ blue, azure blue, violet, deep mauve.

Riebeckite and magnesio-riebeckite:

α prussian blue, bluish green;
β indigo blue, bluish green, pale violet;
γ yellowish green, brownish yellow.

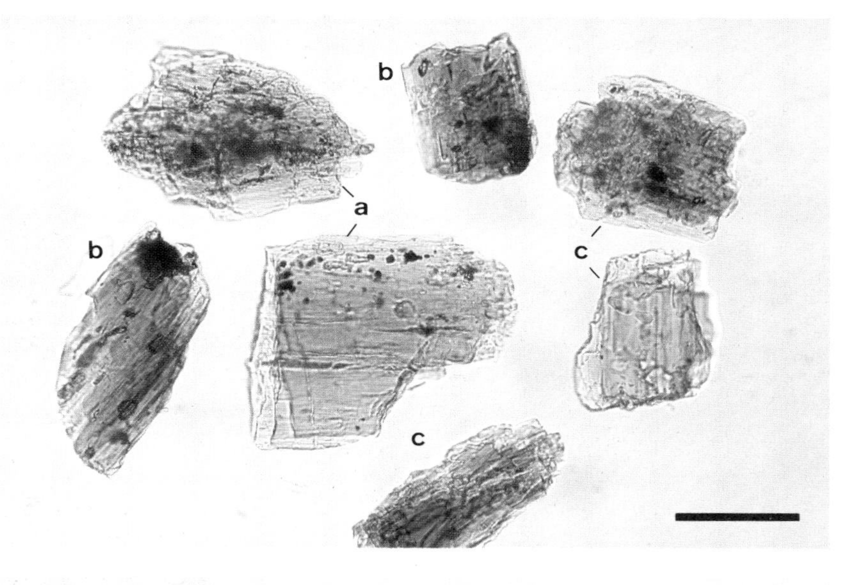

The shade of pleochroism may vary within a grain. Because of dominant {110} orientation of the grains upon mounting, α is rarely discernible.

Birefringence: Closely related to composition and is usually moderate, but increases slightly with higher Mg-content. Interference colours of the paler varieties are in the range of first-order grey or yellow, but those of the deeper coloured specimens are usually masked by natural mineral colour. Patchy and uneven polarization colours are fairly common.

Extinction: The extinction angle measured on (010) ranges between 4° and 14° for glaucophane composition and is somewhat higher for riebeckite. Cleavage fragments normally have nearly parallel extinction, which is often mottled owing to compositional variation.

Interference figure: {110} prisms of glaucophane composition yield centred acute bisectrix figures with fairly large 2V, a negative sign and strong dispersion. This figure is clearly visible in grains which show normal interference tints. Strong mineral colour may obscure observation. The interference figures of members of crossite to riebeckite compositions usually consist of poorly defined eccentric isogyres. 2V of crossite is very small but good bisectrix figures are seldom obtainable, therefore diagnosis cannot be made on this basis.

Elongation: Due to compositional variation, positive and negative elongations may alternate within a single grain. Glaucophane is length slow, but that which is close to **ferro-glaucophane** composition may be length fast. Grains of crossite, magnesio-riebeckite and riebeckite are length fast.

Distinguishing features: Blue colour and characteristic pleochroism, prismatic or occasionally fibrous morphology, and amphibole cleavages ensure an easy diagnosis of the series itself. Paler tints, weaker pleochroism and lack of colour in the α direction, generally positive elongation and the tendency to yield good acute bisectrix figures, may suffice for the diagnosis of glaucophane. Tints in greenish-blue shades, strong colour in the α direction and negative elongation can indicate riebeckite. However, positive identification of the individual members within the series can be made only by chemical means. Blue tourmaline is uniaxial, has poor cleavages and its greater absorption colour is perpendicular to the vibration direction of the lower nicol. Arfvedsonite displays very strong dispersion and usually abnormal interference colours. The relief of chloritoid is higher and it has a mica-like habit. Blue dumortierite exhibits stronger pleochroism in indigo-blue shades.

Occurrence: Glaucophane and crossite are characteristic minerals of the glaucophane schist and glaucophanitic greenschist facies in which magnesio-riebeckite may also form. They are the so-called high-pressure index minerals, which crystallize together with lawsonite, pumpellyite, jadeite, etc. They are frequently found with the same association in sediments. Glaucophane is also present in garnet–glaucophane amphibolites. Riebeckite crystallizes in a range of environments from late diagenetic to medium-grade metamorphism. It is common in alkali igneous rocks. In some metamorphosed iron formations it forms together with magnesio-riebeckite but they are also present in non-metamorphosed iron formations. Authigenic riebeckite has been reported from the Green River Formation of Utah (USA) by Milton & Eugster (1959).

Remarks: When chemical analysis is not carried out, it is customary to refer to members of the glaucophane–riebeckite series as **blue sodic amphiboles**.

Grains from: (a) Glaucophanitic amphibole, Oligocene Molasse, Savoy, France; (b) ferro-glaucophane, Oligocene Molasse, Savoy, France; (c) crossite, Burdigalian Molasse, Savoy, France (Mmt 1.582).

Arfvedsonite

monoclinic, biaxial (−)

$n\alpha$	1.674–1.700
$n\beta$	1.679–1.709
$n\gamma$	1.686–1.710
δ	0.005–0.012
Δ	3.00–3.50

Form in sediments: Grains are generally prismatic fragments which show well developed prismatic cleavages. Fibrous forms may also occur. Grains of pyroclastic origin sometimes poikilitically enclose feldspars and apatite.

Colour: Arfvedsonite is highly coloured in shades of brown and bluish green. Grains may be very dark blue, brown or almost non-transparent in the γ direction.

Pleochroism: Strong pleochroism with varying absorption is characteristic of arfvedsonite: α, deep bluish green, indigo, yellow, dark green; β, green, yellowish brown, grey, violet; γ, pale yellowish green, deep green, brownish green.

Birefringence: Weak to moderate. Interference colours are obscured by deep mineral colour and absorption, hence grains show extreme variability of shades and spectacular colour changes from yellowish brown and brown to vivid green on stage rotation.

Extinction: Owing to strong dispersion the extinction is often incomplete and is variable even within a single grain. $\alpha : z = 5°–30°$ for arfvedsonite and 20°–35° for magnesio-arfvedsonite.

Interference figure: Some cleavage flakes yield flash figures with variable 2V. However, most grains exhibit eccentric dark interference figures because of strong absorption and mineral colour. $v>r$ dispersion of the optic axes is strong and determination of the optic sign is difficult.

Elongation: Negative.

Distinguishing features: Amphibole habit and cleavages, strong colour, coupled with marked pleochroism, as well as incomplete extinction, are diagnostic. Glaucophane is length slow and pleochroic in shades of blue and violet. Arfvedsonite may be confused with some length-fast blue amphiboles, especially riebeckite which has a similar pleochroic scheme. However, the pleochroism of arfvedsonite always includes a brownish colour, absent in most blue amphiboles. Riebeckite has a smaller extinction angle and may be optically positive. Eckermannite is pleochroic in blue green and yellow green and its extinction angle is larger. Blue-green pargasitic, edenitic or magnesio-hornblendes have weaker pleochroism, higher-order interference tints and positive elongation. Blue tourmaline is uniaxial and its maximum absorption is perpendicular to the vibration direction of the polarizer. The extinction of dumortierite is parallel, it shows higher-order polarization colours and a different pleochroic scheme.

Occurrence: Arfvedsonite is a characteristic constituent of plutonic alkali igneous rocks, mainly granites, syenites, nepheline-syenites and their associated pegmatites. It occurs rarely in alkali-rich volcanics. Milton *et al.* (1974) reported

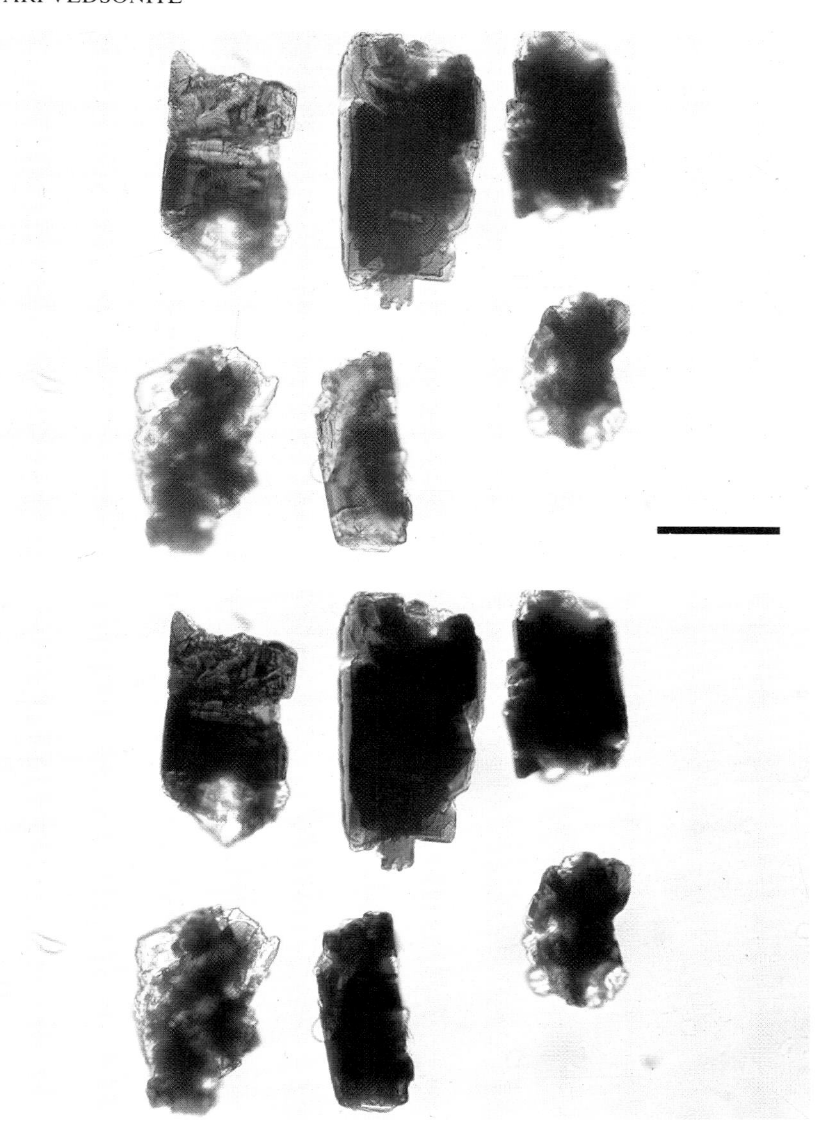

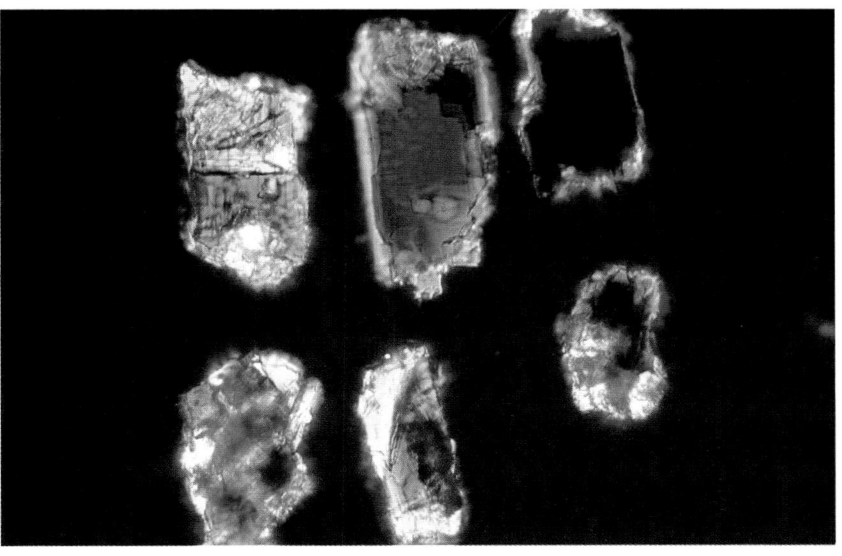

authigenic magnesio-arfvedsonite from the Green River Formation of Utah (USA).

Grains from: Pyroclastic origin from the Palaeogene Thanet Formation, borehole 79/7A 57.89 m, offshore south-east England (Mmt 1.582). Courtesy of A. C. Morton.

MICA GROUP

Muscovite

(including other white micas)

$K_2Al_4[Si_6Al_2O_{20}](OH,F)_4$
monoclinic, biaxial (−)

$n\alpha$	1.552–1.574
$n\beta$	1.582–1.610
$n\gamma$	1.587–1.616
δ	0.036–0.049
Δ	2.77–2.88

Form in sediments: Muscovite appears invariably as {001} basal plates, usually with a round, more rarely with an irregular, outline. The plates may be bent, curled or have fine-grained fibrous or scaly muscovite on their margins. Inclusions of magnetite, hematite and accessory minerals are often present.

Colour: Colourless, highly transparent, but may be stained yellow.

Pleochroism: Non-pleochroic.

Birefringence: Basal plates have weak birefringence resulting in first-order pale grey or white interference colours which range up to pale yellow in thicker plates. Grains lying perpendicular to the cleavage have strong birefringence and second- or third-order interference colours.

Extinction: Nearly parallel to cleavage traces. Basal plates often show wavy or mottled extinction.

Interference figure: Basal plates provide excellent, centred acute bisectrix figures with well-defined isogyres in a white or yellow field. One or two isochromatic curves may appear, usually at the outer part of the figure. 2V is rather variable depending on the composition and structure. r>v dispersion is distinct.

Elongation: Positive to cleavage traces.

Distinguishing features: Mica habit and lack of colour, together with an excellent interference figure, provide an easy diagnosis. Talc resembles muscovite but is usually composed of fine small overlapping scales, lacks complete extinction and is virtually uniaxial.

Occurrence: Musvocite is present in a wide variety of metamorphic rocks and is especially common in schists and gneisses. In plutonic rocks it forms in granites, pegmatites, aplites and in hydrothermal veins. Sericite is produced by the hydrothermal alteration of feldspars of igneous and metamorphic rocks and also during metamorphism of intermediate and acid rocks.

Remarks: **Paragonite** and **phengite** are optically indistinguishable from muscovite and it is common to use the term '**white mica**' for the colourless mica varieties when further optical distinction is not possible. The name **sericite** is used to describe fibrous, scaly, fine-grained, usually secondary mica.

MUSCOVITE

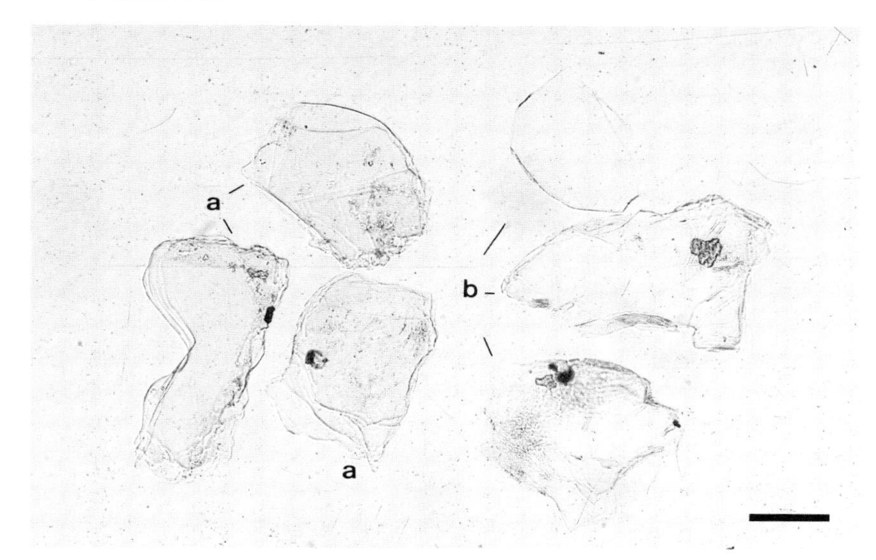

Grains from: (a) River sand, Dartmoor, England; (b) Carboniferous, North Sea (Mmt 1.539).

Glauconite

$(K,Na,Ca)_{1.2-2.0}(Fe^{3+},Al,Fe^{2+},Mg)_{4.0}[Si_{7-7.6}Al_{1-0.4}O_{20}]$
$(OH)_4.n(H_2O)$
monoclinic, biaxial (−)

$n\alpha$	1.592–1.610
$n\beta = \gamma$	1.614–1.641
δ	0.014–0.030
Δ	2.40–2.95

Form in sediments: Glauconite is found in both the heavy and light mineral fractions. It displays a considerable variation in morphology and of physical and chemical properties. Grains are granules, pellets or infillings of predominantly carbonate microfossil tests, but irregular fragments and glauconitized mineral grains are also common. Glauconite is usually cryptocrystalline. It has a perfect cleavage which is rarely visible on detrital grains.

Colour and pleochroism: Thicker grains are very dark green or almost opaque. Small pellets and thinner fragments show different shades of green, sometimes bluish or yellowish-green colours, and are weakly pleochroic in green and yellow.

Birefringence: Moderate to strong. Glauconite has aggregate polarization with second-order green and yellow interference tints. These are often obscured by strong mineral colour and are best seen with the substage condenser inserted.

Extinction: Parallel.

Interference figure: Owing to the fine crystalline structure of the aggregates, detrital grains seldom yield interference figures.

Elongation: Positive.

Distinguishing features: Colour, morphology and cryptocrystalline structure are distinctive. Chlorites may appear in aggregate form similar to glauconite, but their birefringence is very low. The mode of occurrence of **berthierine** is similar to that of glauconite and they are indistinguishable megascopically or by optical means. X-ray diffraction often reveals that grains formerly identified as glauconite are berthierine. Celadonite is similar to glauconite minerals, but its chemical composition is different and it is often associated with volcanic materials. Some pumpellyite aggregates may resemble glauconite but their relief is markedly higher.

Occurrence: Glauconite and berthierine are restricted to the marine environment and are formed by the replacement of mineralogically different substrates. They are common constituents of greensands, glauconitic sandstones, glauconitic limestones and marls. Glauconite encountered in sediments was either formed *'in situ'* or was recycled from pre-existing glauconitic sediments.

Remarks: Odin & Matter (1981) proposed to discontinue the term 'glauconite' and to use instead the term **glaucony** for the facies and **glauconitic smectite** or **glauconitic mica** for the minerals. It appears, however, that these terms have not yet come into general practice and the name glauconite is still in use. To overcome the problem some authors refer to 'glauconitic materials'.

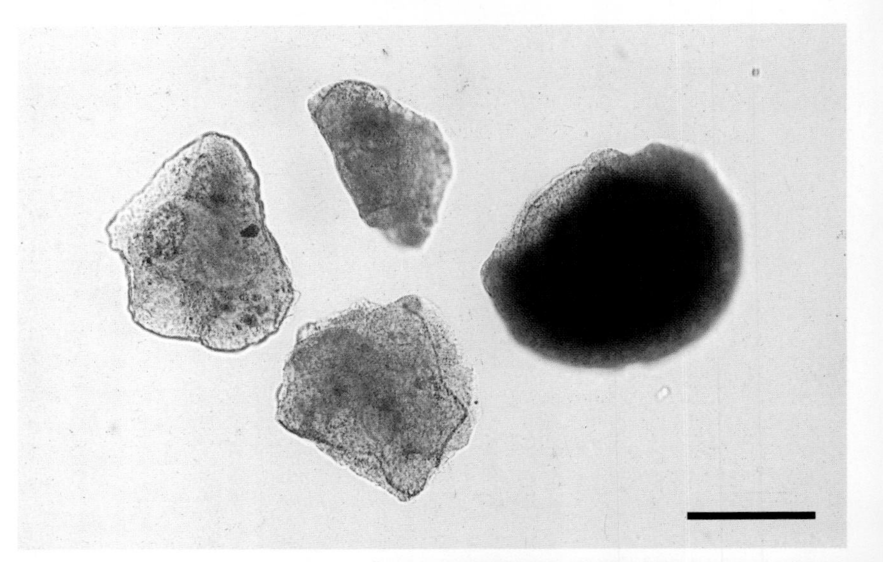

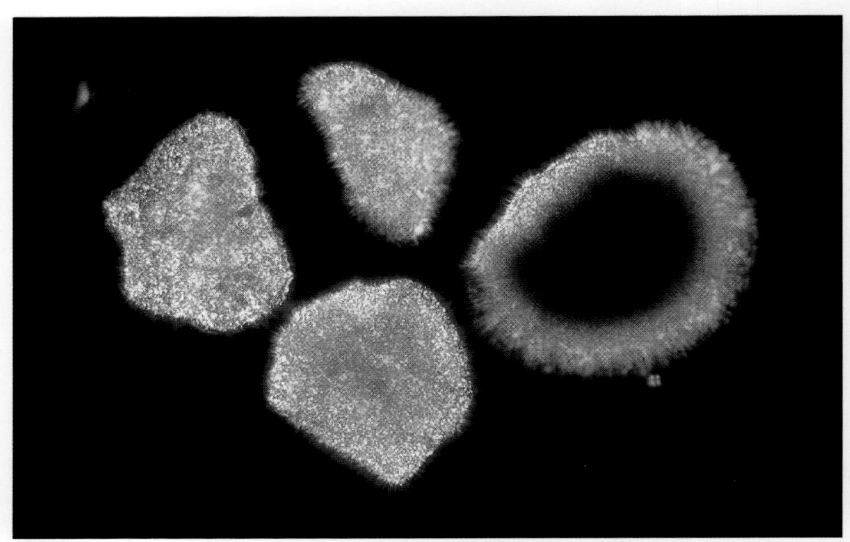

Grains from: Oligocene, Nerthe-chain, France (Mmt 1.582).

Biotite

$K_2(Mg,Fe^{2+})_{6-4}(Fe^{3+},Al,Ti)_{0-2}[Si_{6-5}Al_{2-3}O_{20}]O_{0-2}(OH,F)$
monoclinic, biaxial (−)

$n\alpha$	1.565–1.625
$n\beta = \gamma$	1.605–1.696
δ	0.04–0.08
Δ	2.70–3.30

Form in sediments: Grains are platy with an irregular or round outline and, on mounting, commonly settle on their basal plane. Thin flakes, foliated or shredded particles are occasionally encountered. Biotites of pyroclastic origin are predominantly euhedral with a pseudo-hexagonal habit (a, c). Inclusions of zircon, apatite, sphene and opaque minerals are common.

Colour: Various shades of reddish brown, brown and greenish brown are characteristic. Green biotite (lower) may be quite common in sediments but is often mistaken for chlorite. Leaching of biotite in sediments produces very pale brown or pale yellow colours. Chloritization is manifested in a gradual change of brown colours to green, accompanied by significantly lower-order interference colours, even within a single grain.

Pleochroism: Basal sections are weakly pleochroic or non-pleochroic, but other orientations show distinct pleochroism in pale green, pale brown or pale yellow to deep green, deep brown, red or dark brown.

Birefringence: Basal plates have weak birefringence and low-order white or yellow interference colours, but these are often masked by strong natural colour. Sections normal to the basal cleavage have strong birefringence and second-order orange, red and green interference colours.

Extinction: Almost parallel (3°–9°). Mottling, close to the extinction position is characteristic (lower, right).

Interference figure: Basal cleavage flakes yield centred acute bisectrix figures with small 2V. The isogyres usually fail to show separation, thus the interference figure appears uniaxial.

Elongation: To the cleavage traces is positive.

Distinguishing features: Flaky or foliated mica habit, or occasionally pseudohexagonal form, colours and very small 2V are chief diagnostic features. Brown amphiboles are dominantly prismatic, show amphibole cleavage and stronger birefringence. Non-pleochroic basal sections of brown tourmaline with low-order interference colours may be confused with biotite, but the former lacks cleavages or foliation. Green biotite has higher birefringence than chlorite.

Source rocks: Biotite is ubiquitous in all types of igneous rocks and is especially characteristic of granites and granitic pegmatites. It also occurs in rhyolites and andesites. Biotite is widespread in gneisses, greenschists and in amphibolite facies rocks.

Remarks: **Phlogopite** and **stilpnomelane** are indistinguishable from biotites in sediments. Phlogopite has a fairly

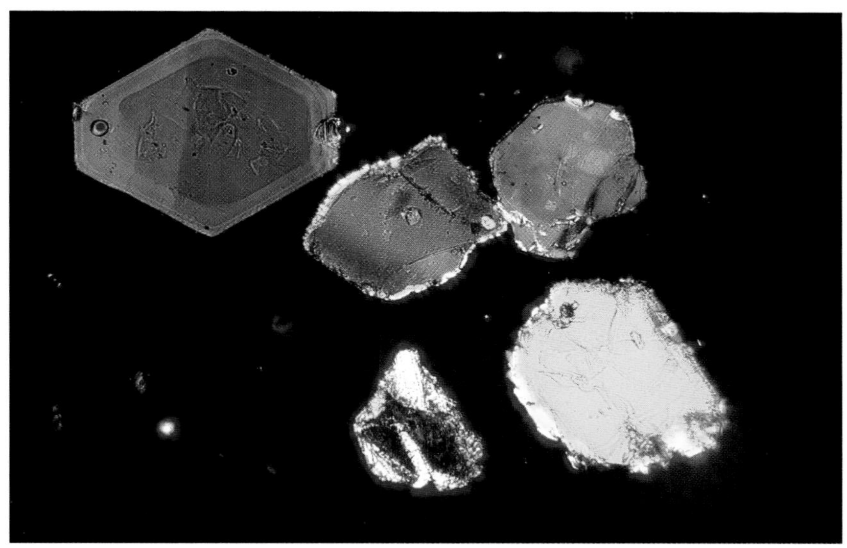

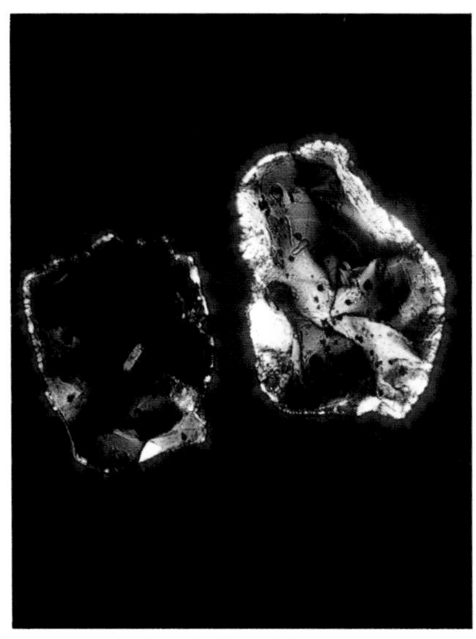

restricted paragenesis, but stilpnomelane is the index mineral of the low- to medium-grade metamorphism (stilpnomelane zone); therefore its detrital occurrence may furnish valuable clues to source terrains. The basal cleavage of stilpnomelane is less perfect than that of biotite. In samples containing fibrous or imperfect flakes of brown or green micas, stilpnomelane may be present. In this instance diagnosis requires chemical tests.

Grains from: Upper: (a) Carboniferous, borehole Weiach, 1393 m, Switzerland; (b) Middle America Trench, DSDP Leg 66; (c) Oligocene, Barrême Basin, France (Mmt 1.582).
Lower: left-hand grain: Tertiary, North Sea; right: Lower Miocene Molasse, Savoy, France (Mmt 1.582).

Talc

$Mg_6[Si_8O_{20}](OH)_4$
monoclinic, biaxial (−)

$n\alpha$	1.539–1.550
$n\beta$	1.589–1.594
$n\gamma$	1.589–1.600
δ	0.05
Δ	2.58–2.83

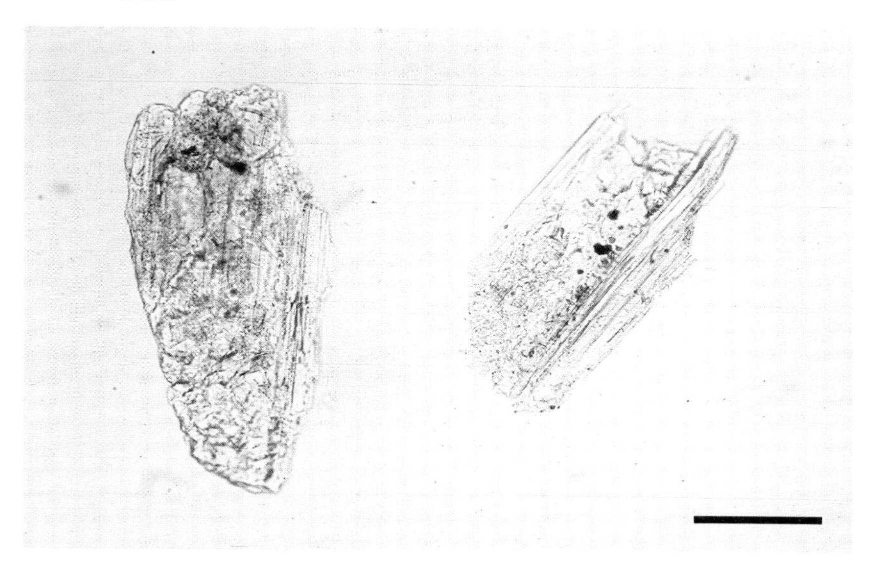

Form in sediments: Talc appears as mica-like round, oblong or irregular {001} basal plates, either as thin single sheets or as composite plates of small overlapping scales. It is often foliated or seen as aggregates of fine needles or fibres. It may contain opaque inclusions.

Colour: Essentially colourless, but staining can cause yellowish or brownish tints.

Pleochroism: Non-pleochroic.

Birefringence: Basal plates have low birefringence and display dull first-order grey, white or pale-yellow interference colours. Fibrous and composite grains of randomly oriented flakes have strong birefringence associated with second- and third-order yellow, orange and blue interference colours.

Extinction: Either wavy or undulatory, because of superimposed fine flakes, but most often grains fail to extinguish in any position.

Interference figure: Basal plates, even composite grains of fine scales yield a well-centred acute bisectrix figure with very small 2V. The clear isogyres in a white or yellow field hardly separate and the figure appears uniaxial. Isochromatic bands are usually present in thicker grains.

Elongation: Oblong flakes and fibres have positive elongation.

Distinguishing features: Single basal plates of talc greatly resemble white mica, from which distinction can be made by the virtually uniaxial figure of the former. Composite platelets or overlapping small flakes are characteristic of talc, but are rarely found in white micas. It is difficult to distinguish sericite from fibrous talc. Talc resembles pyrophyllite, but the latter is unlikely to be found in clastic sediments. Talc may be a common detrital species, but is probably mistaken for sericite or white mica.

Occurrence: Talc forms by the hydrothermal alteration of ultrabasic rocks, producing talc-rich schists ('steatitization'). In metamorphic rocks it is found in talc–anthophyllite, talc–termolite and talc–antigorite schists and also in thermally metamorphosed siliceous dolomites. In evaporites it may form authigenically.

Grains from: Soil sample from an unknown locality in Africa, found together with magnesio-anthophyllite (Mmt 1.539).

Chlorite group

General formula: $(Mg,Al,Fe)_{12}[(Si,Al)_8O_{20}](OH)_{16}$
monoclinic, biaxial (+) (−)

$n\alpha$	1.57–1.66
$n\beta$	1.57–1.67
$n\gamma$	1.57–1.66
δ	0–0.01
Δ	2.60–3.30

Though within the chlorite group several species are distinguished on a structural and compositional basis, in grain mounts identification of a specific member cannot be made with certainty. Chlorites may be detrital or are formed authigenically during diagenesis. They sometimes dominate the heavy mineral fractions but in general practice are excluded from grain counts.

Form in sediments: Detrital specimens are dominantly thin flaky basal cleavage plates of round, oval or irregular shape, often with curled margins. Some grains may be shredded. Scaly aggregates or fibres also occur. Especially authigenic grains are of the latter types. Rulite, zircon, apatite and opaque inclusions are frequent.

Colour: Detrital chlorites exhibit various shades of green, sometimes in a patchy arrangement.

Pleochroism: Not always visible, but deeply coloured varieties usually show distinct pleochroism from pale green, yellowish green to deep bluish green or greenish brown.

Birefringence: Basal plates have near zero birefringence and appear dull dark grey or almost isotropic under crossed polars. Iron-rich varieties generally exhibit anomalous indigo-blue, sometimes purple, interference tints. Those of non-basal sections or variously oriented fibres display second-order yellow and orange colours.

Extinction: Almost parallel to cleavage traces and to fibres. Birefringent basal plates usually have a wavy or mottled extinction.

Interference figure: Basal plates may yield off-centre or well-centred biaxial figures with varying 2V in a bluish field.

Elongation: Fibrous and shredded grains have either positive or negative elongation.

Distinguishing features: Mica habit, green colour and low birefringence are distinctive. Green biotite is likely to be mistaken for chlorite, but the former has considerably higher birefringence. The relief of flaky green serpentines is lower, they lack pleochroism and have higher birefringence. Glauconite is granular or pelletal and has higher RI. Pumpellyite may be confused with chlorite but its relief and birefringence are stronger.

Occurrence: Chlorite is widespread in low-grade metamorphic rocks and is most common in the greenschist facies. In igneous rocks, chlorite is generated by the hydrothermal alteration of ferromagnesian minerals. Weathering processes can also produce chlorite and in sedimentary rocks it often forms authigenically during diagenesis. Detrital grains of

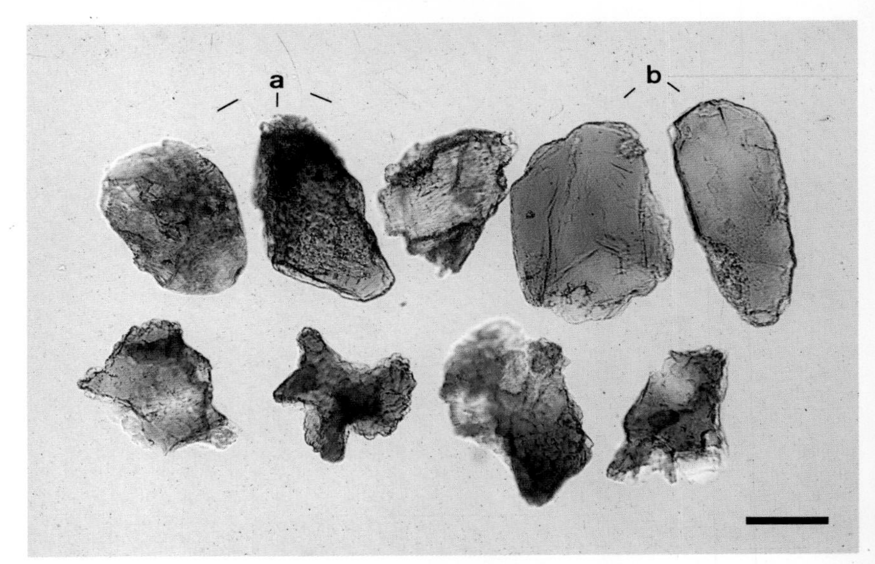

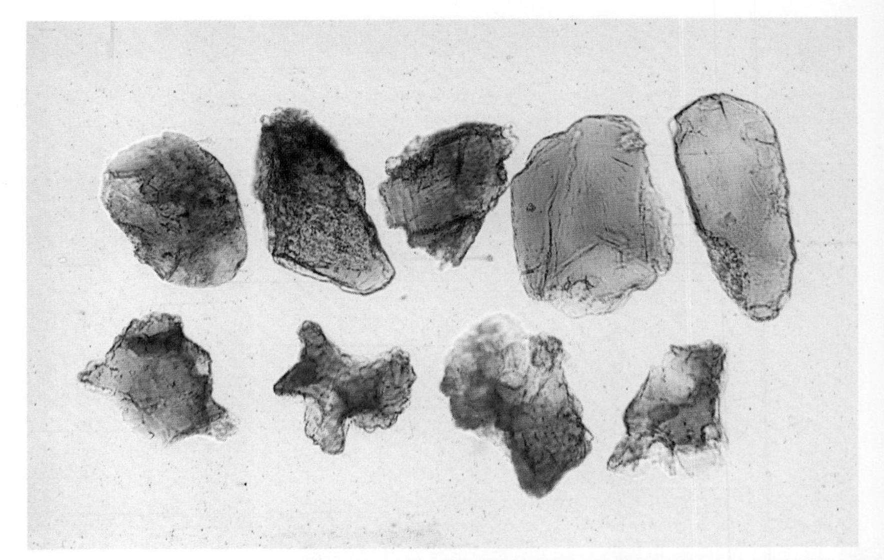

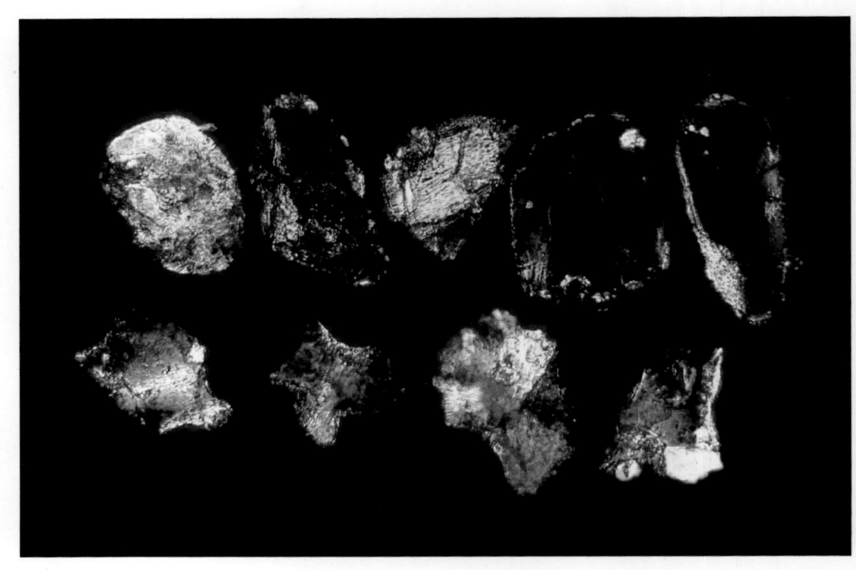

consolidated rocks may be coated with a thin film of authigenic chlorite.

Grains from: (a) Lower Miocene Molasse, Savoy, France; (b) Oligocene, Barrême Basin, France.
Lower row: authigenic chlorites, Rotliegend, northern Germany (Mmt 1.539).

Serpentine group

$Mg_3[Si_2O_5](OH)_4$
monoclinic, biaxial (−)

	chrysotile	lizardite	antigorite
$n\alpha$	1.532–1.549	1.538–1.554	1.558–1.567
$n\beta$	1.530–1.564	-	1.560–1.573
$n\gamma$	1.545–1.556	1.546–1.560	1.562–1.574
δ	0.013–0.017	0.006–0.008	0.004–0.007
Δ	≤2.55	≃2.55	2.60

The above three principal polymorphic varieties of the serpentine group are usually designated as serpentine when encountered in sediments, because optical properties alone are inadequate for distinguishing between the three species. Grains may be regarded as serpentine minerals or fragments of serpentinites. Their density (~2.6) suggests that they belong to the light mineral fraction, but because of the common presence of iron ore inclusions they tend to be concentrated in the heavy mineral residue.

Form in sediments: Serpentine generally occurs as platy grains with a low relief and a smooth round or oblong outline. Irregular, jagged plates, foliated flakes or fibrous aggregates have also been found. Some grains may exhibit the shape of the serpentinized mineral or contain relics of pyroxenes or olivine. As a by-product of serpentinization, iron ore inclusions are extremely common, and often only a thin rim around a seemingly opaque grain indicates the presence of a serpentine mineral. Grains appear translucent with a greasy or pearly lustre under the binocular microscope.

Colour: Thin flakes are almost colourless, very pale yellow or pale green, but shades deepen to light brown or green with increasing thickness.

Pleochroism: Rarely noticeable. Some darker-coloured varieties exhibit very faint pleochroism.

Birefringence: Low. Thin plates appear pale, or dark grey or white, under crossed polars. With stronger natural colour and increasing thickness, interference colours range to second-order yellow, orange, green and sometimes anomalous bluish green.

Extinction: Wavy, undulatory or mottled, owing to differing orientation of platelets or fibres within a single grain.

Interference figure: It is very difficult to obtain an interference figure. Larger homogeneous areas of some grains may show flash figures.

Elongation: Positive.

Distinguishing features: Low relief is characteristic and the outline of the thinner grains nearly disappears in a low-index medium (e.g. Canada balsam). Flaky or fibrous habit and the very common presence of opaque inclusions are diagnostic. The lower relief together with higher-order interference colours help to distinguish serpentine from micas, especially from chlorite. Leached biotites may resemble serpentine but their RI is higher and they yield good interference figures.

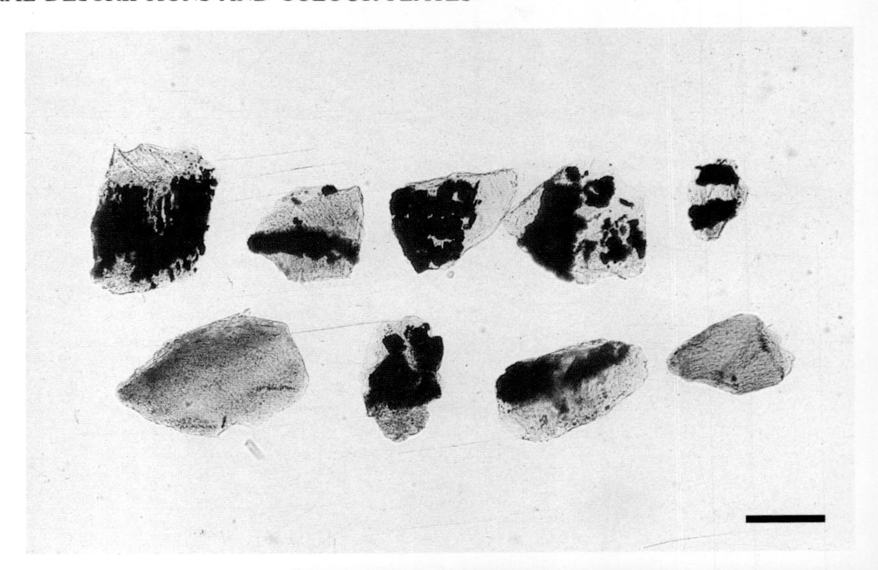

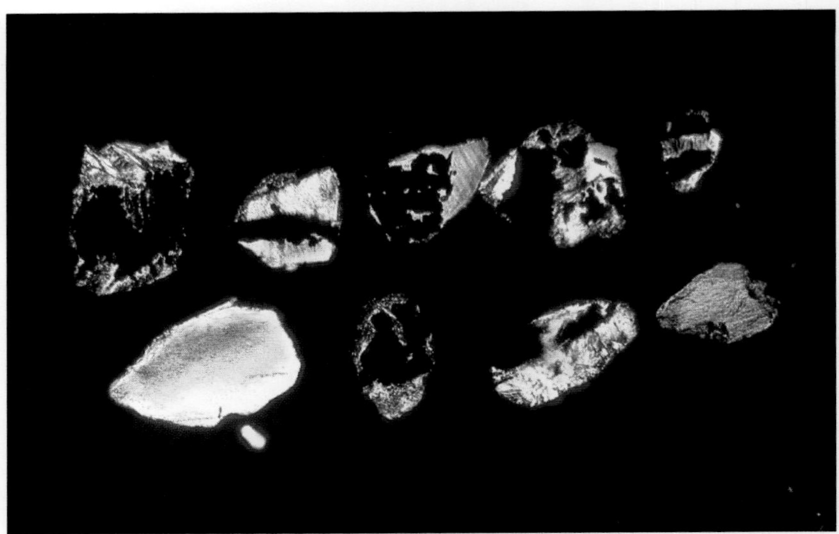

Occurrence: Serpentine minerals are formed essentially by the serpentinization of ferromagnesian minerals in ultrabasic rocks. They occur, but less commonly, in metamorphic rocks and in hydrothermal veins.

Grains from: Oligocene, Barrême Basin, France (Mmt 1.582).

Prehnite

$Ca_2Al[AlSi_3O_{10}](OH)_2$
orthorhombic, biaxial (+)

$n\alpha$ 1.611–1.632
$n\beta$ 1.615–1.642
$n\gamma$ 1.632–1.665
δ 0.022–0.035
Δ 2.90–2.95

Form in sediments: Grains are most frequently basal-cleavage fragments which are often composed of small irregular or triangular-shaped sectors. These may be portions of larger crystals exhibiting characteristic 'bow-tie' structure. Sheaf-like aggregates and clusters of radially arranged fibres may also be present. The surface of the grains is usually flaky. Sometimes fine lamellar twinning is discernible. Inclusions are minute fluid globules or microlites.

Colour: Prehnite is colourless, but frequent iron oxide staining produces a yellowish tinge.

Pleochroism: Non-pleochroic.

Birefringence: Moderate. Interference colours are upper first- to lower second-order shades, usually dominated by yellow and orange. Some grains may exhibit abnormal polarization colours.

Extinction: Parallel to cleavage traces, but often wavy or incomplete.

Interference figure: Obscured by the fibrous or composite nature of the grains.

Elongation: Either positive or negative.

Distinguishing features: Prehnite is recognized by fibrous or platy habit and moderate relief. The presence of small sectors within a grain is particularly characteristic. Wavy extinction, lamellar twinning and flaky surface are also typical. It can be mistaken for colourless pumpellyite, but the latter has pronounced anomalous interference colours.

Occurrence: Prehnite is a diagnostic mineral of low-grade regionally metamorphosed prehnite–pumpellyite facies rocks, where it often forms veins. It occurs in a similar manner in meta-greywackes. Prehnite is common in basic volcanic rocks, in some hydrothermal veins and in contact-metamorphic impure carbonate rocks.

Remarks: As far as we are aware, prehnite has not been previously recorded as a detrital mineral. It was identified in the sandstones of the Burdigalian Upper Marine Molasse of Switzerland at several locations. It occurs together with pumpellyite in a fairly diverse first-cycle mineral assemblage (Mange-Rajetzky & Oberhänsli 1986), and is also present in the Oligocene of the Barrême Basin, France. As prehnite is a common constituent of several rock types, occurring often on a broad scale, it will probably be found in sediments elsewhere.

Grains from: Grain lower right: Pleistocene fluvial deposit of the River Arve, western Switzerland; Other grains: Lower Miocene Molasse, Yverdon, Switzerland (Mmt 1.582).

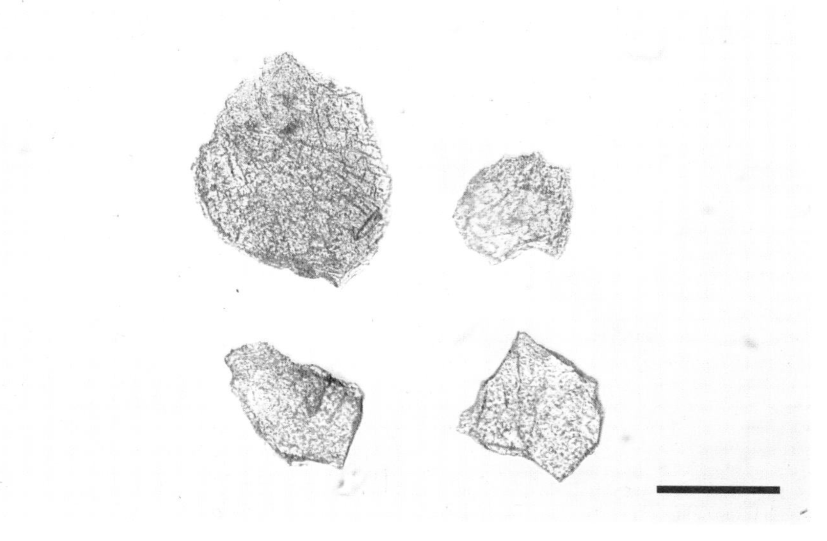

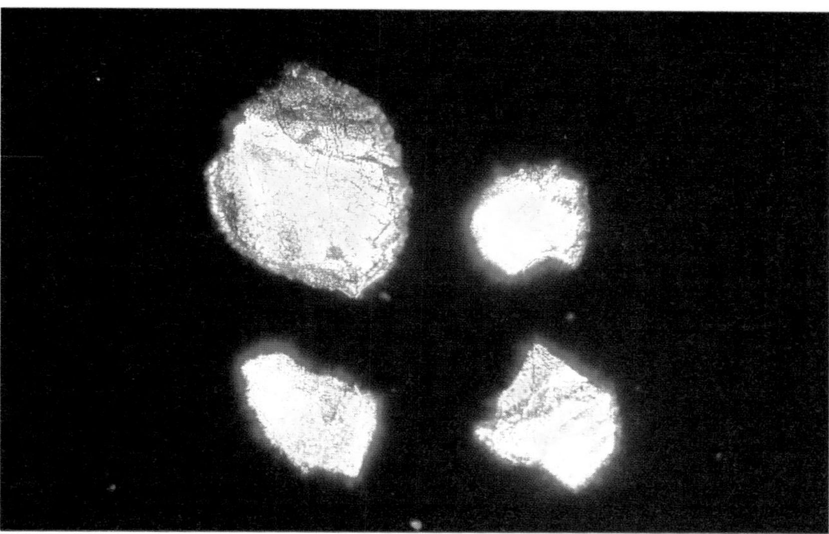

OXIDES

Cassiterite

SnO_2
tetragonal, uniaxial (+)

$n\omega$	1.990–2.010
$n\varepsilon$	2.091–2.100
δ	0.096–0.098
Δ	6.98–7.02

Form in sediments: Detrital cassiterites have a diversity of forms and they occur as euhedral crystals, prismatic or sharp irregular fragments, usually with an uneven rough surface, as well as subrounded to rounded grains. Conchoidal fractures and striae, parallel to the principal axis or to pyramidal faces, are typical features. Some grains may appear 'dusky'.

Colour: Extreme relief and adamantine or submetallic lustre are characteristic. It exhibits a wide range of colours from colourless to pale yellow, golden yellow, pink, red, deep brownish red, brown and almost opaque. Irregular colour patterns are frequent, which appear either as colour zoning or as deep-coloured and pleochroic patches alternating with colourless or almost opaque portions within a single grain.

Pleochroism: Varies through very weak to strong and ranges from colourless to deep red, pink or yellow.

Birefringence: Very strong. Interference colours are exceedingly high and bright, but may be masked by the strong inherent mineral colour.

Extinction: Euhedral crystals and prismatic fragments have parallel extinction.

Interference figure: The majority of detrital grains are anomalously biaxial and exhibit an interference figure with widely spaced isogyres on a white field. Some fragments, which show weaker interference colours, yield an off-centre uniaxial figure with dense isochromatic colour bands. The determination of the optical sign is difficult.

Elongation: Positive, but difficult to observe.

Distinguishing features: Extreme relief and birefringence, adamantine lustre, deep colour together with often irregularly distributed colour patches are diagnostic. Prismatic rutile strongly resembles cassiterite, but the former has a higher relief, stronger birefringence and has an even colour distribution. Cassiterite can be confused with prismatic fragments of zircon, although the latter has lower RI and birefringence, as well as homogeneous colours. Brookite and sphene are biaxial positive and show intense, often anomalous, interference tints. The birefringence of anatase is low and it is uniaxial negative. Sphalerite resembles cassiterite but it is isotropic.

Occurrence: Cassiterite is a constituent of highly siliceous igneous rocks. In granites, hydrothermal processes generate cassiterite, but sometimes it is a primary accessory of granites. In greisen, granite pegmatites and in hypothermal veins, cassiterite is a common constituent. Wood-tin is formed by low-temperature colloidal precipitation. Cassiterite is highly

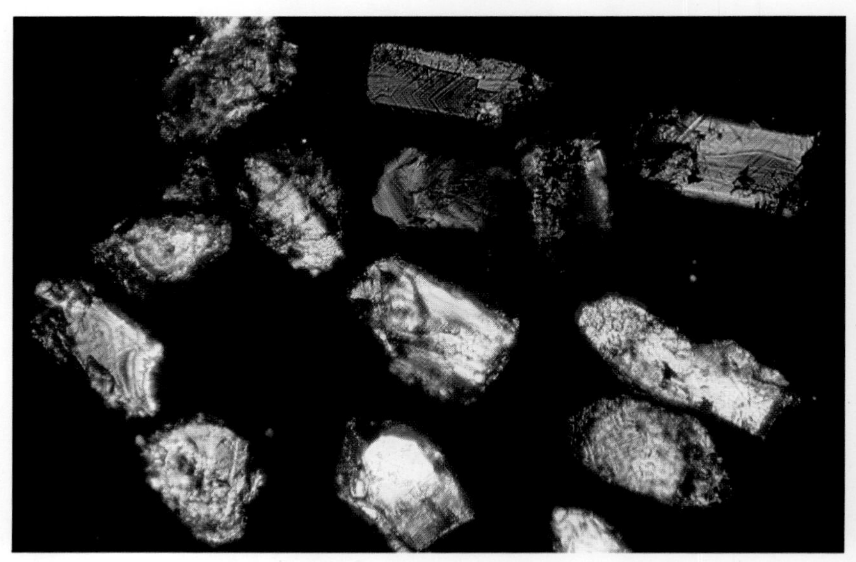

resistant to weathering (Hails 1976).

Grains from: Left of dividing line: Kinta Valley tin field, Malaya; Right of line: Beach sand, St Ives, Cornwall, England (Mmt 1.662).

Corundum

Al_2O_3
trigonal, uniaxial (−)

$n\omega$ 1.767–1.772
$n\varepsilon$ 1.760–1.763
δ 0.008–0.009
Δ 3.98–4.02

Form in sediments: Grains are generally irregular, angular, subrounded and occasionally well rounded. Basal parting produces fragments with a hexagonal outline (left). Conchoidal fractures and small indentations are often exhibited.

Colour: Detrital corundum appears in shades of blue, but colourless grains have also been reported. Inhomogeneous colour distribution is characteristic.

Pleochroism: The intensity of pleochroism varies from weak to very strong: ω, pale to deep blue; ε, colourless or pale bluish green. A combination of ω deep blue and reddish pink may appear within one grain.

Birefringence: Weak, but detrital grains (unless they are very thin fragments) exhibit anomalously high interference colours of third order. The sometimes intensive mineral colour may mask the interference tints.

Extinction: Elongated fragments and parting plates have parallel extinction.

Interference figure: Basal parting plates provide a good uniaxial interference figure on a white field. Irregular grains often yield anomalously biaxial interference figures with large 2V.

Elongation: Negative.

Distinguishing features: High relief, blue colour, often in a patchy distribution, and pleochroism are characteristic. High-order interference colours exhibited by corundum grains are inconsistent with the low birefringence of this mineral and this is also typical for corundum. Apatite has lower RI and birefringence; saphirine is biaxial, has a strong dispersion and is length slow. The relief of anatase is higher, its shape is dominantly rectangular and it shows very low-order interference colours.

Occurrence: Corundum is present in a variety of rock types but is most characteristic of Al-rich, Si-poor rocks. It forms in syenites and their pegmatitic equivalents and also in quartz-free pegmatites which transect ultramafic rocks. Contact- and regionally metamorphosed rocks, especially those derived from aluminous or carbonate sediments, contain corundum and it may also form in mica or chlorite schists. Corundum is a common constituent of emery deposits.

Remarks: Carborundum may contaminate samples during preparation and can be confused with corundum. However, the former has a higher relief and appears as sharp, fractured particles, often with a conchoidal pattern and it is optically uniaxial positive.

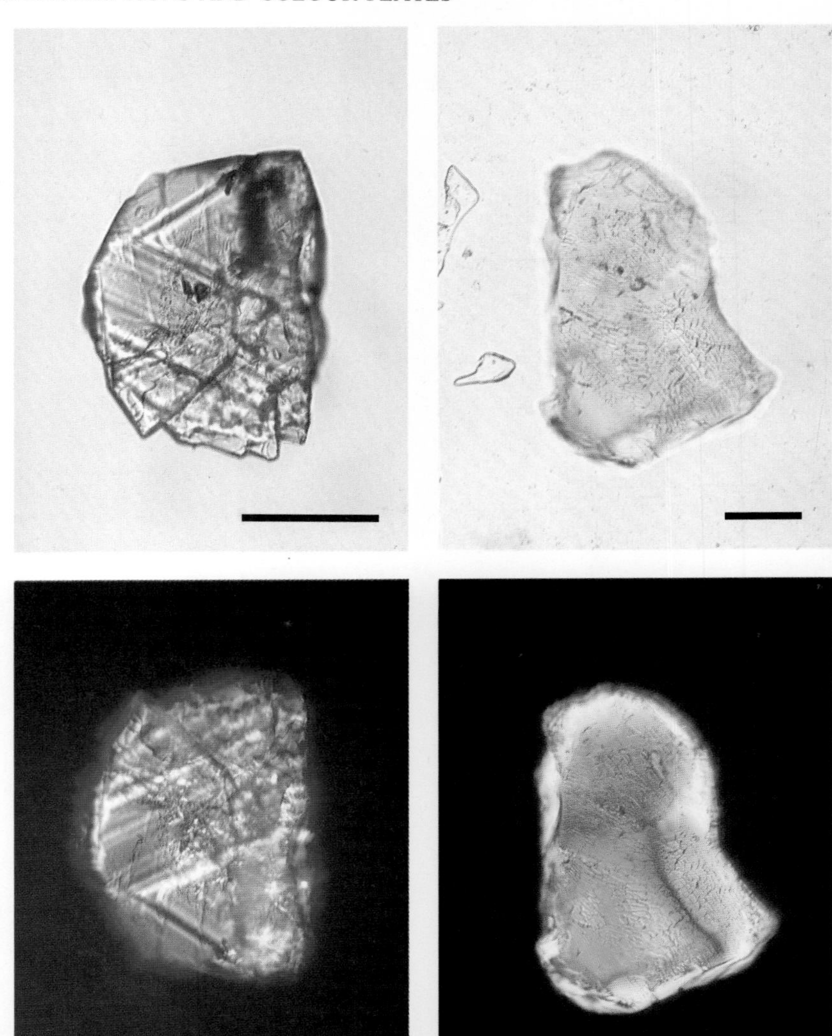

Grains from: Right: Triassic, northern Tunisia (Mmt 1.662); Left: Upper Miocene, western Hungary (CB).

Rutile

TiO_2
tetragonal, uniaxial (+)

$n\omega$	2.605–2.613
$n\varepsilon$	2.899–2.901
δ	0.286–0.296
Δ	4.23–5.50

Form in sediments: Detrital rutiles are generally small grains which rarely exceed 200 μm. Because of extremely high refractive indices, a thick black halo surrounds the grains. Subrounded to well-rounded oblong forms and spheres are the most frequent. Non-rounded grains are either euhedral, with well-developed pyramidal terminations, or slender prisms, prismatic fragments and sometimes fractured grains, showing irregular or conchoidal breakage patterns. 'Knee-shaped' twins (d) and parallel crystal growths are sometimes encountered. Grains may have multi-lamellar twinning often in two directions, arranged diagonally to the prism's face (c), or show striations parallel with their long axis.

Colour: Shades of red, such as deep blood red and brownish red (left) are more frequent in sediments than brownish-black, yellowish-brown, amber or violet-brown colours (right).

Pleochroism: Distinct, and usually manifested in the deepening of the red, yellow or brown shades.

Birefringence: Extreme, but the high-order interference colours are generally obscured by the strong inherent mineral colour. Grains usually display reddish-white, red, brownish or yellowish interference tints.

Extinction: Prismatic and oval-shaped grains show parallel extinction.

Interference figure: Thinner fragments occasionally yield well-centred uniaxial figures with many isochromatic bands in a red or yellow field. However, many grains have biaxial characters which may be due to twinning or deformation.

Elongation: Owing to strong mineral colour and birefringence, detection of the optical orientation is difficult. The positive elongation may be observed on paler-coloured varieties.

Distinguishing features: Rutile is diagnosed by its very high relief, deep colours, pleochroism and extreme birefringence. Striations or twinning, when present, are also characteristic. The only mineral with which rutile may be confused is cassiterite, but the RI of the latter is lower, its colour often shows patchy arrangement, and many grains have a 'dusky' appearance. Translucent hematite is flaky, and brookite shows abnormal interference colours.

Occurrence: Rutile is a widespread accessory mineral in metamorphic rocks, particularly in schists, gneisses and amphibolites. It is less significant in igneous rocks, where it occurs in hornblende-rich plutonic types and in pegmatites. In a survey of rutile parageneses Force (1980) concluded that the source rocks of detrital rutile are predominantly those of high-grade regionally metamorphosed terrains.

Rutile has been reported to crystallize authigenically in

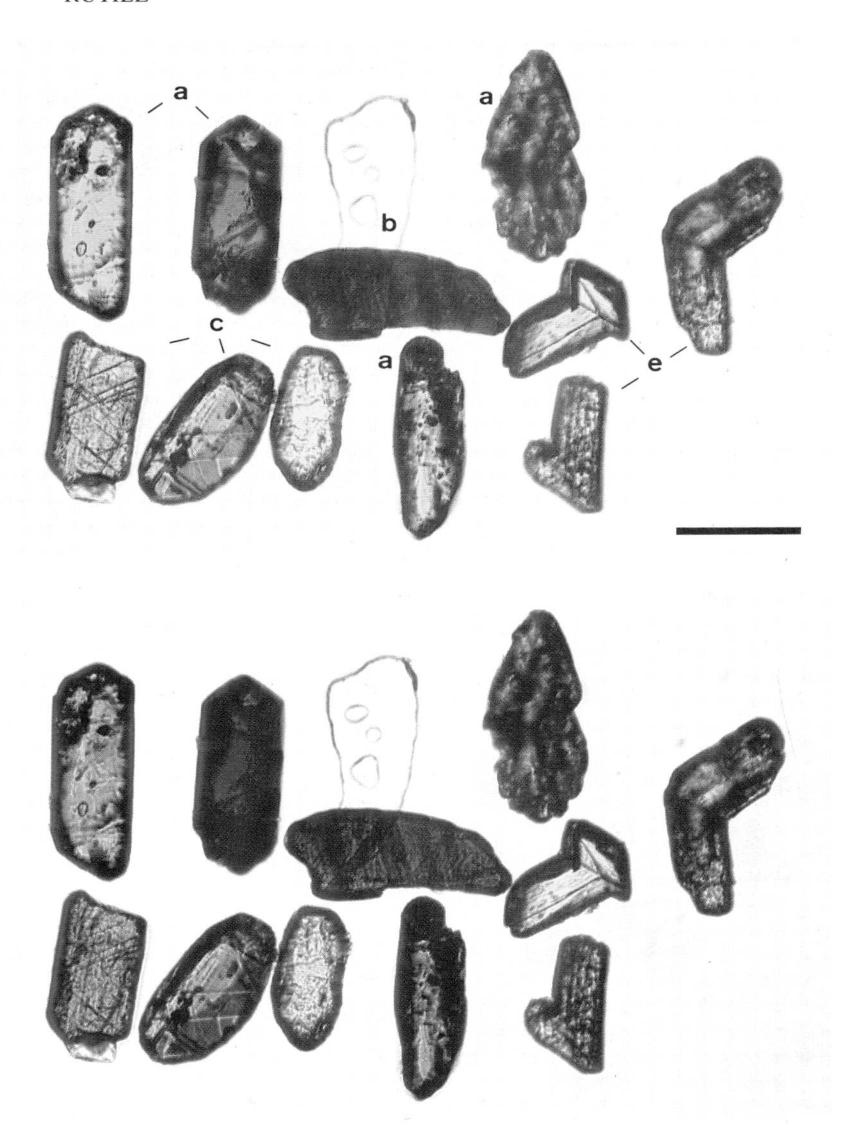

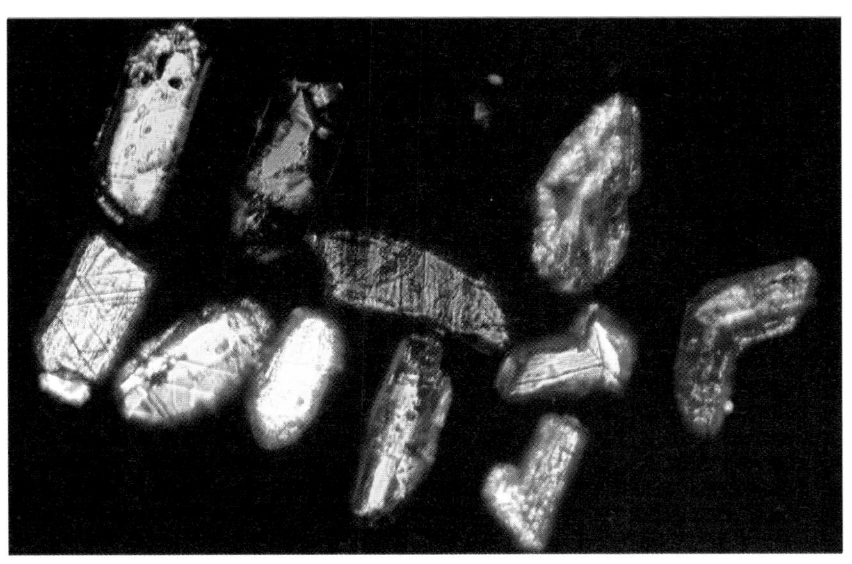

sediments in which it forms clusters of needles or thin laths, often intergrown with other Ti-bearing (leucoxene, anatase, sphene) phases. These varieties can be distinguished easily from the detrital grains by the different morphology and considerably smaller size.

Remarks: Rutile is an ultrastable mineral and is one of the three index species (zircon–tourmaline–rutile (ZTR), Hubert 1962), which are used to characterize the mineralogical maturity of a heavy mineral suite. The presence of well-rounded rutiles indicates recycled sedimentary source rocks.

Grains from: Left page: (a) and (c) Tertiary, North Sea; (b) and (e) Jurassic, North Sea; (d) Lower Miocene, Nerthe-chain, southern France (Mmt 1.662).
Right page: Grab sample, Port of Beira, Mozambique (Mmt 1.662).

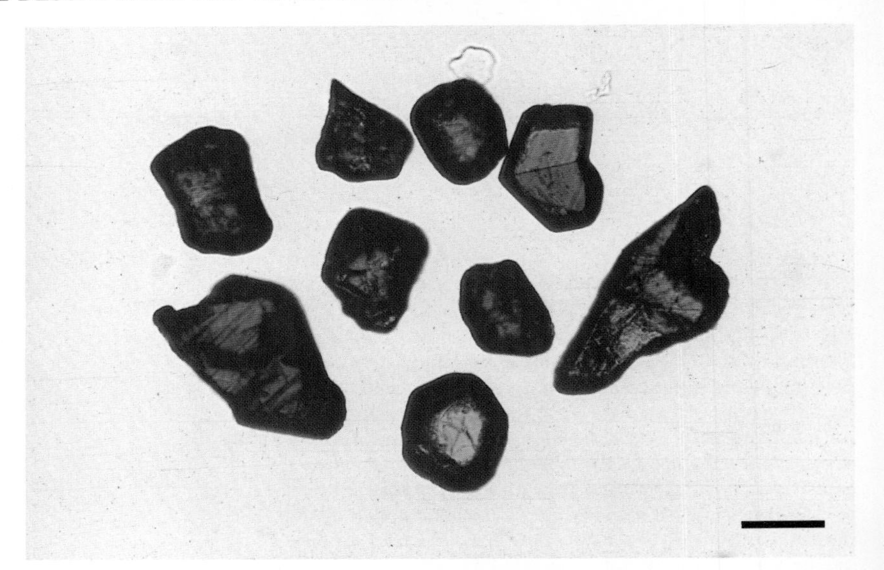

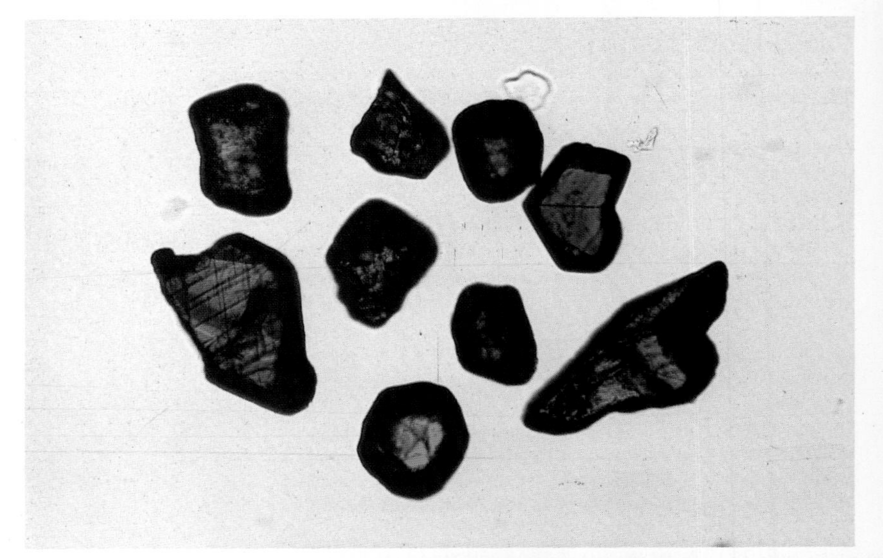

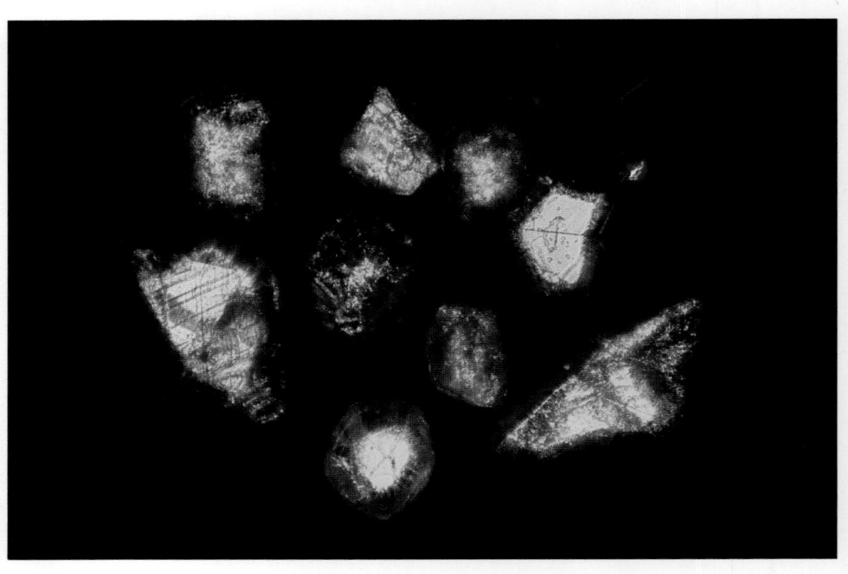

Anatase

TiO_2
tetragonal, uniaxial (−)

$n\omega$	2.561
$n\varepsilon$	2.488
δ	0.073
Δ	3.82–3.97

Form in sediments: Anatase has an extreme relief, submetallic lustre and it occurs both as detrital grains and as authigenic crystals. The authigenic forms are dominantly tabular, rectangular basal plates (upper). Several of these may be superimposed (b) or are intergrown, forming a composite grain (c). Zoning or geometric patterns are frequently displayed. Anatase also appears as small, almost opaque aggregates. Detrital grains (lower), though they are uncommon, may be encountered as rounded basal plates or rather dull, dusky irregularly shaped grains. Occasionally pyramidal or octahedral (lower, top left) specimens appear which often have striated faces.

Colour: Shades of blue, yellow and pale brown; colourless grains are uncommon.

Pleochroism: Deep-coloured thick anatase may display pleochroism: ω, pale yellow, pink or blue; ε, blue, dark blue or orange.

Birefringence: Non-basal plates have strong birefringence with interference tints of fourth-order white, grey and pink. Basal plates have yellow, faint dark grey, blue or brown polarization colours or are non-birefringent.

Extinction: Parallel or symmetrical to cleavage traces and crystal faces.

Interference figure: Basal plates with faint interference colours usually provide good uniaxial interference figures. Some varieties may yield biaxial flash figures.

Elongation: Negative.

Distinguishing features: Extreme relief, habit and colours are diagnostic. Distinction between authigenic and detrital anatase is made chiefly by the morphology of the grains. Brookite is biaxial positive and almost always yields a well-centred acute bisectrix figure. The RI of spinel is lower and is isotropic. Corundum has a lower relief, different morphology and brighter interference colours. Rutile is uniaxial positive and exhibits very strong birefringence.

Occurrence: In igneous and metamorphic rocks anatase is the low-temperature polymorph of TiO_2. It is common in sediments and forms predominantly authigenically at the expense of unstable Ti-bearing minerals. It is a constituent of leucoxene. Detrital anatase grains are sourced principally by igneous or metamorphic rocks as the thin fragile authigenic crystals of sediments are unlikely to survive recycling.

Grains from: Upper: (a) and (b) Rotliegend, northern Germany; (c) Tertiary, Saudi Arabia; (d) Buntsandstein, Triassic, Mumpf, Switzerland (Mmt 1.662).

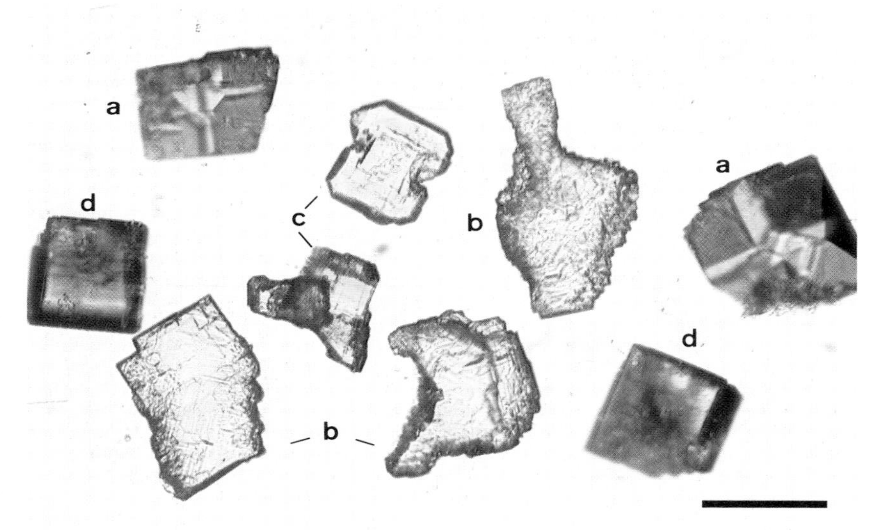

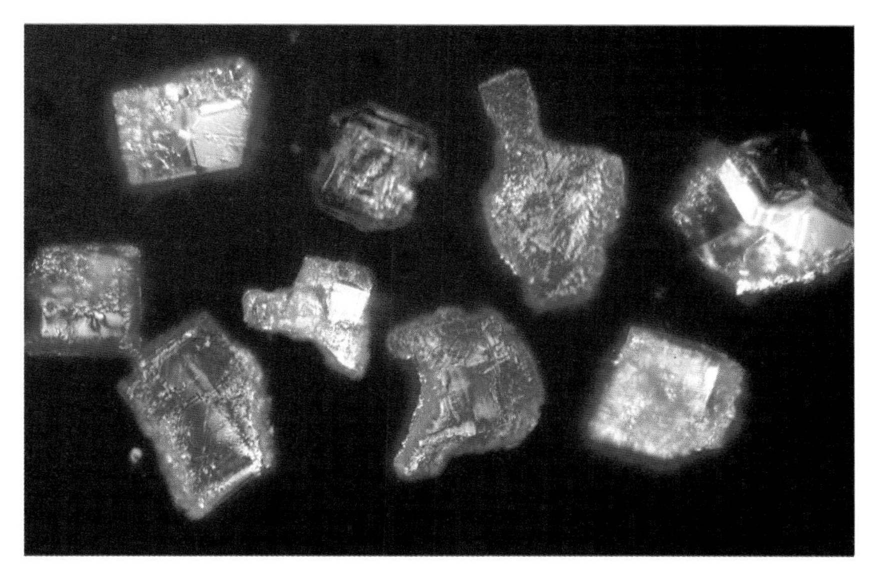

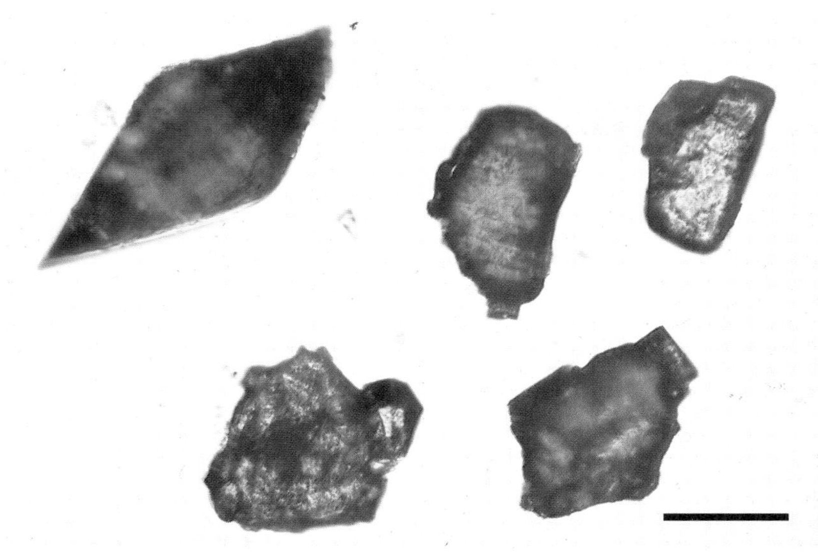

Lower: grains top left: river sand, Val Scareglia, Switzerland; Other grains: Buntsandstein, Triassic, borehole Weiach, 982 m, Switzerland (Mmt 1.662).

Brookite

TiO_2
orthorhombic, biaxial (+)

$n\alpha$	2.583
$n\beta$	2.584–2.586
$n\gamma$	2.700–2.741
δ	0.117–0.158
Δ	4.08–4.18

Form in sediments: Brookite has a very high relief and occurs as tabular or irregularly shaped grains, sometimes with a curved outline. Prismatic faces may exhibit striae parallel to the *c* axis. Brookite also appears as somewhat 'dusky' brownish-yellow aggregates (lower).

Colour: Yellowish brown, brown, sometimes reddish.

Pleochroism: The weak yellow, orange and brown shades of pleochroism are rarely visible on detrital grains.

Birefringence: Extreme high birefringence is distinctive. Interference colours are abnormal because of strong axial plane dispersion. Spectacular colours change on stage rotation from yellow, green or blue violet to reddish pink or orange. Aggregates have high-order but duller colours, and the colour change is not always detectable.

Extinction: Brookite usually fails to extinguish in any position. When extinction is visible, it is mostly parallel to the direction of striations.

Interference figure: Many grains provide a centred acute bisectric figure with strong crossed axial-plane dispersion. Isochromatic bands vary from many to none.

Elongation: Negative, but rarely discernible because of masking effects by intensive polarization colours.

Distinguishing features: Colour, extreme relief and birefringence, combined with strong dispersion and incomplete extinction, are distinctive and they assist in an easy diagnosis of brookite. The recognition of aggregates is more difficult and they can be confused with anatase or denoted as unknown. Some rutile resembles brookite, but the former usually has a prismatic or well-rounded morphology, shows homogeneous interference colours and is uniaxial. The relief of sphene is lower and its colours are paler than those of brookite.

Occurrence: Brookite is an accessory mineral of altered igneous rocks, schists, gneisses and hydrothermal veins. In sediments it is dominantly authigenic.

Grains from: Upper: (a) River sand, Val Scareglia, Switzerland; (b) Oligocene, Barrême Basin, France; (c) Jurassic, North Sea (Mmt 1.662).
Lower: Buntsandstein, Triassic, borehole Kaisten, 102 m, Switzerland (Mmt 1.662).

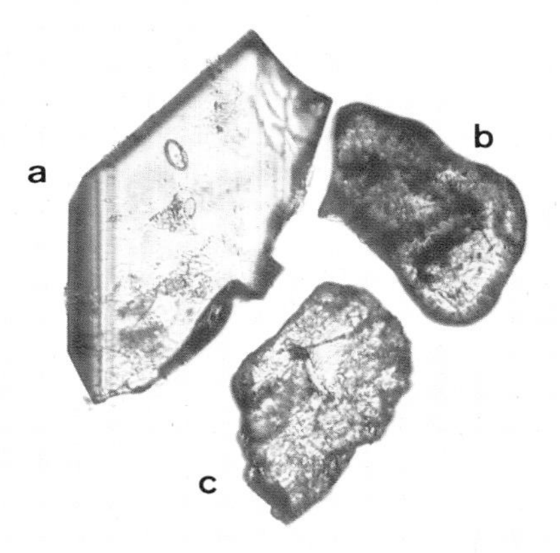

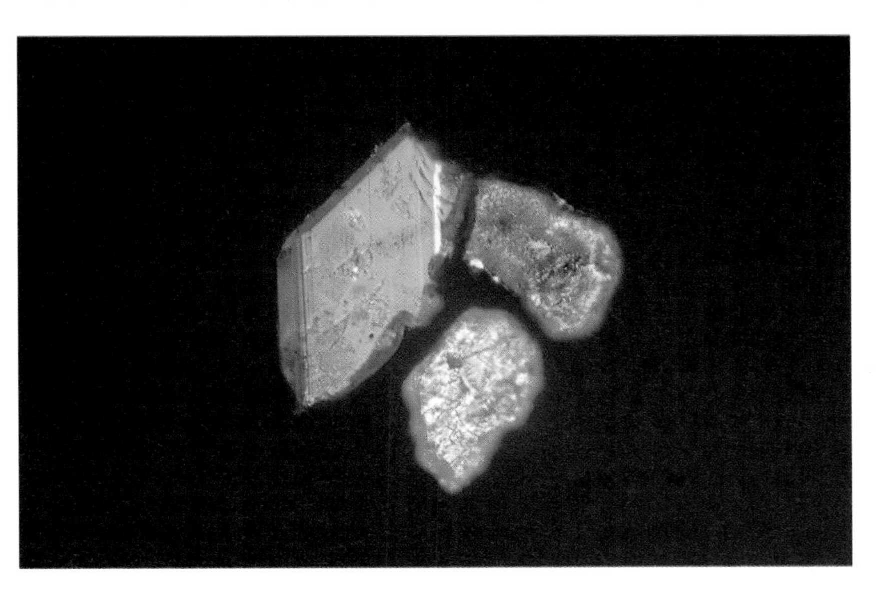

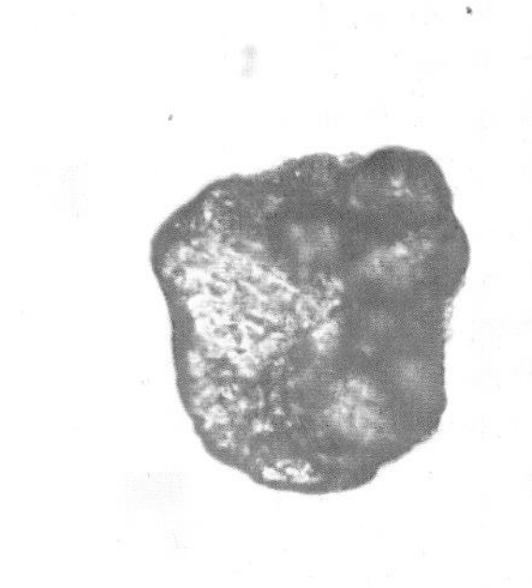

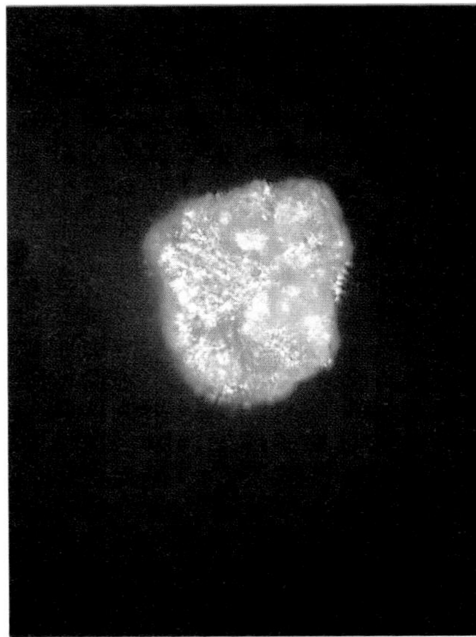

Spinel group

Spinel Series

$(Mg,Fe^{2+},Zn,Mn)Al_2O_4$
isometric

The only widespread species of the spinel series as a detrital mineral is the Cr-rich hercynite picote. It is also referred to as chrome spinel, or brown spinel. Other members have isolated occurrences and are only of local importance. Members of the spinel series may be identified by their high relief, colour, morphology and isotropic character.

Chromian spinel: *n*, 2.00. Δ, 5.09 Grains (upper) are dark reddish brown or clove brown, very sharp angular, irregular fragments which frequently show characteristic conchoidal breakage patterns. Thicker grains are almost opaque and only the thin edges exhibit colour. These grains are best seen with the accessory condenser inserted. Chromian spinel is truly isotropic.

Spinel: *n*, 1.719. Δ, 3.55 It appears as green to blue-green or red, rounded octahedral grains. Broken thinner fragments of the green variety resemble green garnet, but they are completely isotropic.

Hercynite: *n*, 1.835. Δ, 4.40 Grains (lower) are red, reddish yellow, blue or green, and are present in sediments either as rounded octahedral grains or as broken fragments with characteristic conchoidal fractures. Red garnet may be confused with red hercynite, but the former often exhibits anisotropism and is usually speckled with inclusions. When sufficient grains are available, X-ray or electron microprobe analyses will confirm identification.

Occurrence: Chromian spinel is a common accessory of ultramafic rocks such as peridotites and serpentinites. Other spinel varieties are confined to high-grade metamorphic aluminous schists, granitic granulites and gneisses, as well as emery deposits and contact-metamorphic rocks. Spinels are resistant to weathering and may accumulate in gem-gravels.

Remarks: Chromian spinel is frequent in sediments sourced by ophiolitic rocks. The geotectonic significance of detrital chromian spinel (brown spinel) was discussed by Zimmerle (1984). The increase of brown spinel in sediments (with or without serpentinite fragments) in orogenic periods is evident in the geological record. For example, it appears commonly during the Cambrian and Ordovician (Caledonian orogeny), during the Devonian and Carboniferous (Variscan orogeny) and during Cretaceous and Tertiary (Alpine orogeny). Brown spinel is stable both mechanically and chemically, which enhances its significance as a marker mineral, assisting in the reconstruction of the palaeogeographic and palaeotectonic position of ophiolite-bearing structural units in orogenic belts.

Grains from: Upper: (a) Oligocene, Barrême Basin, France; (b) Jurassic, North Sea (Mmt 1.662).
Lower: Monazite sand, India (Mmt 1.662).

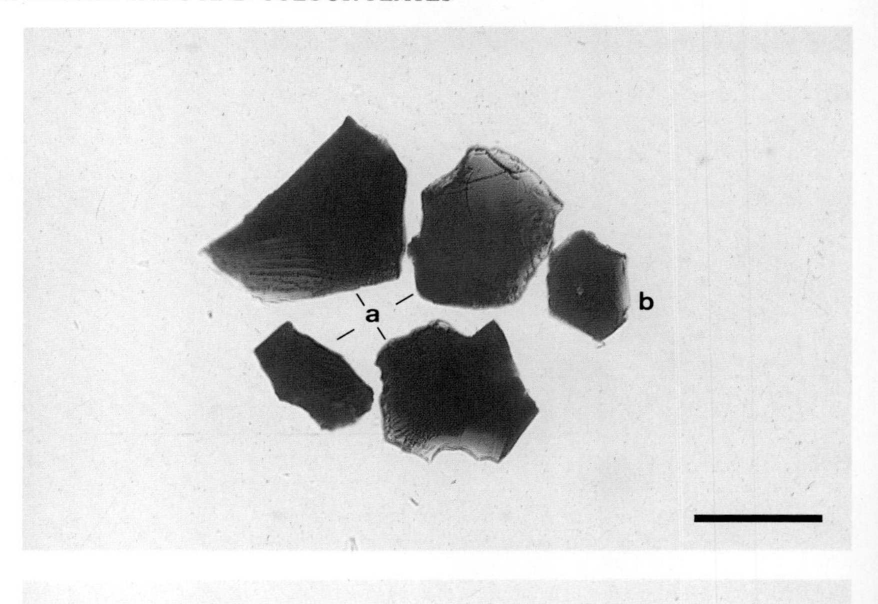

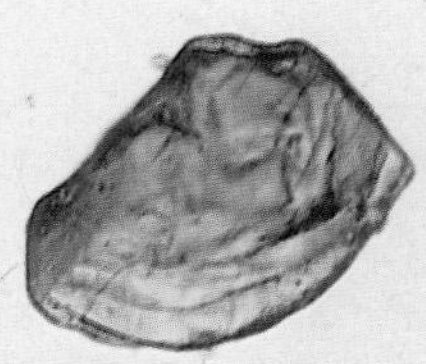

SULPHIDES

Sphalerite

ZnS
isometric

n 2.37–2.50
Δ 4.10

Form in sediments: Grains have a very high relief and are mainly cleavage fragments with an angular irregular form, but tetrahedral, dodecahedral and rounded grains, as well as spherical or granular types, have also been reported (upper). The complete dodecahedral cleavage (in six directions) is manifested in an intricate pattern on the surface of most grains (lower). Opaque impurities may be present. Detrital sphalerite is either clear, translucent or 'dusky', sometimes opaque.

Colour: Shades of yellow, yellowish brown or pale brown are most frequent. On a fractured surface the colour may show an irregular distribution. Colourless grains are also common.

Birefringence: Sphalerite is isotropic but may show weak strain-birefringence.

Distinguishing features: Very high relief, the usual presence of intricate cleavage patterns and isotropic character are distinctive. Cassiterite may have similar relief and colour but has high-order interference tints, whereas sphalerite is isotropic. Anatase has higher RI and a characteristic morphology. The refractive index of garnet is lower.

Occurrence: Sphalerite is common in hydrothermal metalliferous veins, in contact-metamorphic rocks and in replacement deposits in carbonate sediments. It has an affinity to some organic-rich carbonate sediments.

Grains from: Upper: Jurassic, North Sea (Mmt 1.662); Lower: Shore sand, Lelant, Cornwall, England (CB). From the collection of H. B. Milner.

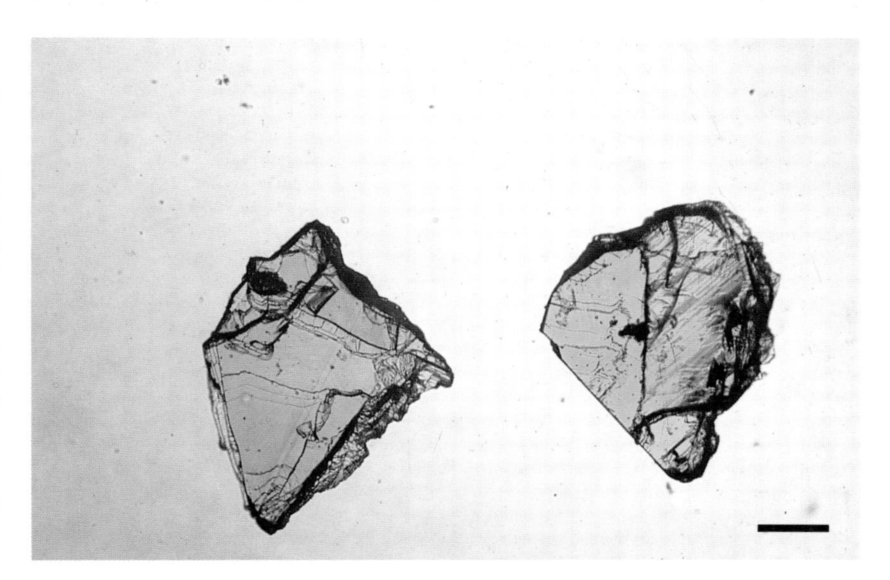

SULPHATES

Baryte

$BaSO_4$
orthorhombic, biaxial (+)

$n\alpha$	1.636–1.637
$n\beta$	1.637–1.639
$n\gamma$	1.647–1.649
δ	0.012
Δ	4.50

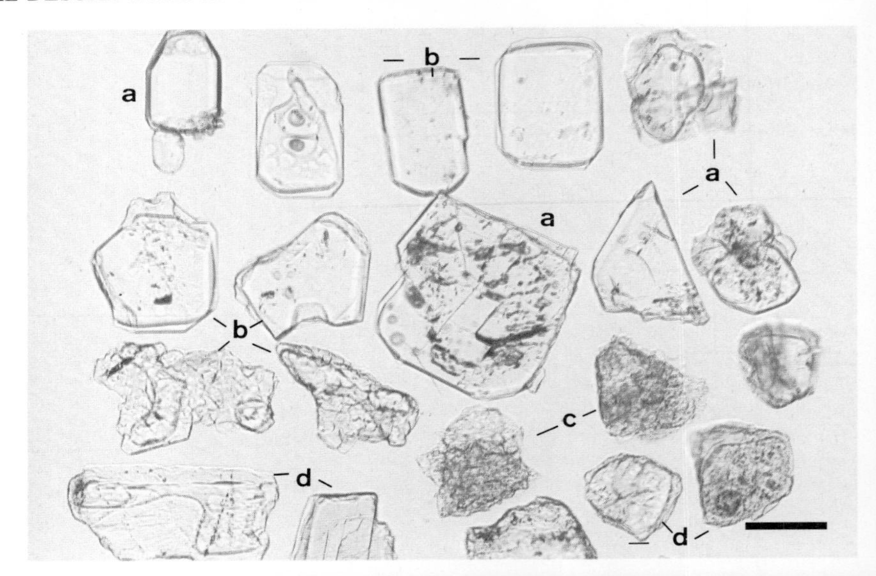

Form in sediments: The commonest habit is sharp irregular or rectangular, the latter produced by perfect {001} cleavage. Euhedral authigenic barytes are usually thin crystals, tabular on {001} and more rarely rhombic prisms (upper, top row). Fractured or granular morphologies may be frequent in heavy mineral suites of baryte-cemented sandstones (lower). Baryte commonly encloses iron oxide, iron hydroxide, sulphide and organic matter and, as a result, grains often appear turbid.

Colour: Colourless, rarely pale yellowish, bluish or grey.

Pleochroism: Seldom observed on detrital grains.

Birefringence: Low, but first-order white, grey or pale-yellow interference colours appear only on thin grains and euhedral crystals laying on the pinacoid. Thicker grains with different orientation show bright yellow, orange and second-order red and greenish-blue polarization colours.

Extinction: Euhedral crystals on the basal pinacoid have parallel extinction, and that of prismatic grains is symmetrical.

Interference figure: Cleavage fragments provide off-centre or flash figures, but grains laying parallel to (100) yield an acute bisectric figure with distinct isogyres and one or two colour bands.

Elongation: Positive on cleavage fragments, but crystals elongated on *x* show negative elongation.

Distinguishing features: Moderate relief, sharp angular form, frequent inclusions and usually bright orange and blue interference colours are distinctive. Cleavage fragments of celestite are likely to be mistaken for barytes, but their RI is somewhat lower. To ensure a positive diagnosis of these two species it is necessary to employ auxiliary analyses, especially for the identification of celestobaryte (see celestite, p. 116). Tabular euhedral barytes may resemble prismatic apatites, particularly misleading being those which display negative elongation. As apatites often show anomalous biaxial positive interference figures, this may enhance confusion. However, apatite is hexagonal and has a higher relief. The birefringence of anhydrite is greater; it has better cleavages and more intense polarization colours.

Occurrence: Baryte occurs in metalliferous hydrothermal veins but may be present in cavities of igneous rocks. Concretions, nodules or veins of barytes are common in carbonate rocks and occasionally in residual clays. Sometimes it is the cementing medium of sandstones.

Remarks: The heavy mineral fractions of drill-cutting

samples are often flooded with the finely ground baryte used as mud additive in drillholes. Grains that lay on the pinacoid and show first-order grey polarization colours greatly resemble detrital apatite, and even experienced analysts find it difficult to distinguish these two minerals. Applying oxalic acid treatment, used for the removal of iron oxide coating, (p. 12) proves useful in these cases. After a few minutes' gentle boiling of the heavy mineral fractions in the solution, apatite grains develop a brownish stain and slight aggregate polarization, which makes identification simple.

Grains from: Upper: (a) Buntsandstein, Triassic, borehole Böttstein, 315 m, Switzerland; (b) Rotliegend, northern Germany; (c) Buntsandstein, Triassic, borehole Berlingen, 2309 m, Switzerland; (d) River sand, Isère, France (Mmt 1.582). Lower: Jurassic, North Sea (CB).

Celestite

$SrSO_4$
orthorhombic, biaxial (+)

$n\alpha$	1.621–1.622
$n\beta$	1.623–1.624
$n\gamma$	1.630–1.631
δ	0.009
Δ	3.96

Form in sediments: Celestite occurs as elongated prismatic fragments, tabular (001) plates (c), irregular particles (d), fibrous fragments (b) or aggregates, composite or 'sheaf'-type forms (a) and less frequently as euhedral crystals. Organic matter may be present and is either distributed irregularly or concentrated in the core of the grains. Minute fluid inclusions are sometimes arranged in parallel strings in two directions, intersecting at almost right angles on the basal plane.

Colour: Colourless, very pale blue or grey, not uncommonly turbid due to impurities.

Pleochroism: Absent on detrital grains.

Birefringence: Weak, and interference colours are first-order grey and white on thin fragments. Thicker grains display higher first-order colours.

Extinction: Basal plates show symmetrical extinction to prismatic cleavages. Prismatic fragments and fibres have parallel extinction.

Interference figure: Basal plates yield an acute bisectrix figure with broad isogyres on a white or yellow field. Cleavage fragments provide eccentric figures.

Elongation: Grains elongated on x have positive elongation and those elongated on y show either negative or positive elongation.

Distinguishing features: Owing to variations of optical and physical properties, a positive identification of celestite grains is difficult. Basal plates, showing moderate relief, symmetrical extinction, low-order interference colours and an acute bisectrix figure are the easiest to diagnose. These grains may lead to the recognition of various other forms usually present in the same sample. The relief of baryte is higher and anhydrite has stronger birefringence.

Celestite is isomorphous with baryte and forms a complete solid solution with it. The name **barytocelestite** or **celestobaryte** is given to strontium-rich barytes.

Occurrence: Celestite occurs predominantly in sedimentary rocks, particularly in carbonates and evaporite deposits, but may also be found as cement of sandstones. Celestite is present in some hydrothermal veins and less commonly as cavity filling in volcanic rocks.

Grains from: (a) Buntsandstein, Triassic, borehole Pfaffnau, 1823 m, Switzerland; (b) Oligocene Molasse, borehole Faucigny-1, 1020 m, France; (c) and (d) Oligocene Molasse, Savoy, France (Mmt 1.582).

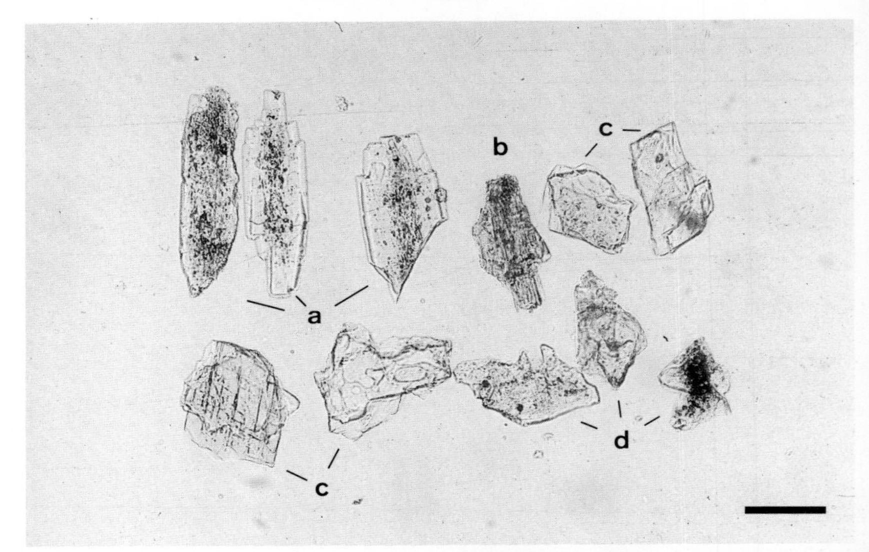

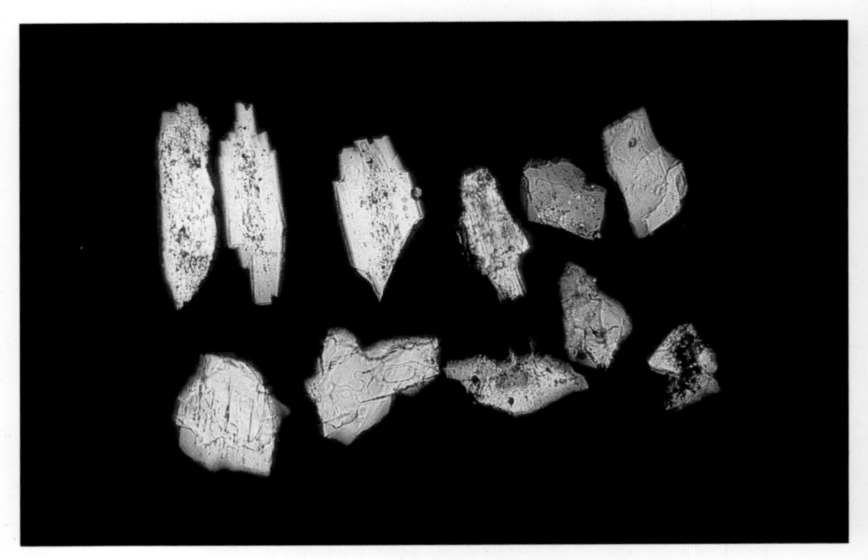

Gypsum

$CaSO_4.2H_2O$
monoclinic, biaxial (+)

$n\alpha$	1.519–1.521
$n\beta$	1.523–1.526
$n\gamma$	1.529–1.531
δ	0.01
Δ	2.30–2.37

Gypsum has a low density and it is retained in the light mineral fraction. Grains containing abundant inclusions may be encountered in the heavy mineral residue.

Form in sediments: The wide range of gypsum habits include euhedral crystals which are usually tabular parallel to {010}, cleavage fragments, diamond-shaped grains, clusters of fibres or fine needles, and spherulitic or granular types. Some grains exhibit simple or contact twins. They frequently enclose inclusions of brine, clays, iron oxide, organic matter, etc. Inclusions may be zonally arranged. Because of the low RI of gypsum, its outline often 'sinks' into the Canada balsam or similar index medium. (Grains are best observed in plane-polarized light, either by oblique illumination or by increasing their relief by partially closing the substage diaphragm.)

Colour: Colourless, sometimes turbid because of numerous inclusions.

Birefringence: Low, and interference colours are in the range of lower or higher first-order tints, depending on the thickness of the grains.

Extinction: {010} tablets have an extinction angle of about 38°. Grains developed according to *b* show parallel extinction.

Interference figure: Grains display poor interference figures. {010} faces yield flash figures. Strong dispersion is rarely visible on detrital species.

Orientation: Elongation is either positive or negative.

Distinguishing features: Very low relief and weak birefringence is characteristic. Grains in the light fraction may be mistaken for thin fragments of quartz, but a Becke test will reveal a higher relief than the Canada balsam for quartz and a lower one for gypsum. The habit of albite and alkali feldspars, their frequent twinning and feldspar cleavages, also the higher β and γ refractive indices of albite, help distinction.

Occurrence: Gypsum is a chemically precipitated mineral, present in a variety of sedimentary environments. Its prominent occurrence is confined to evaporite deposits and it also forms in playa lakes, salt pans and desert soils. Gypsum may be found in some ore deposits and near fumeroles as well as in volcanic vents.

Grains from: Messinian gyps-arenite, Camerino Marche, Italy (Mmt 1.539).

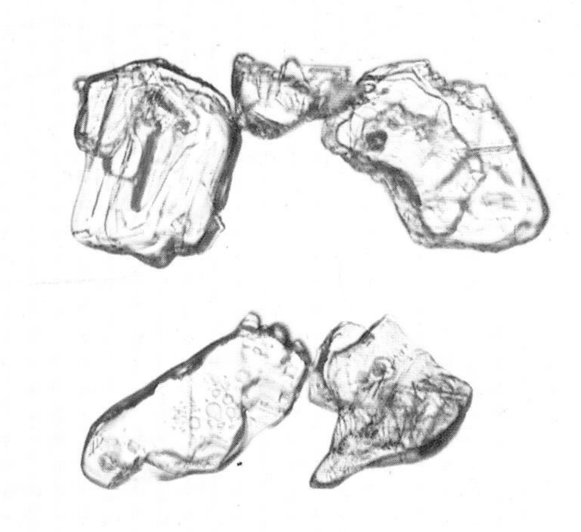

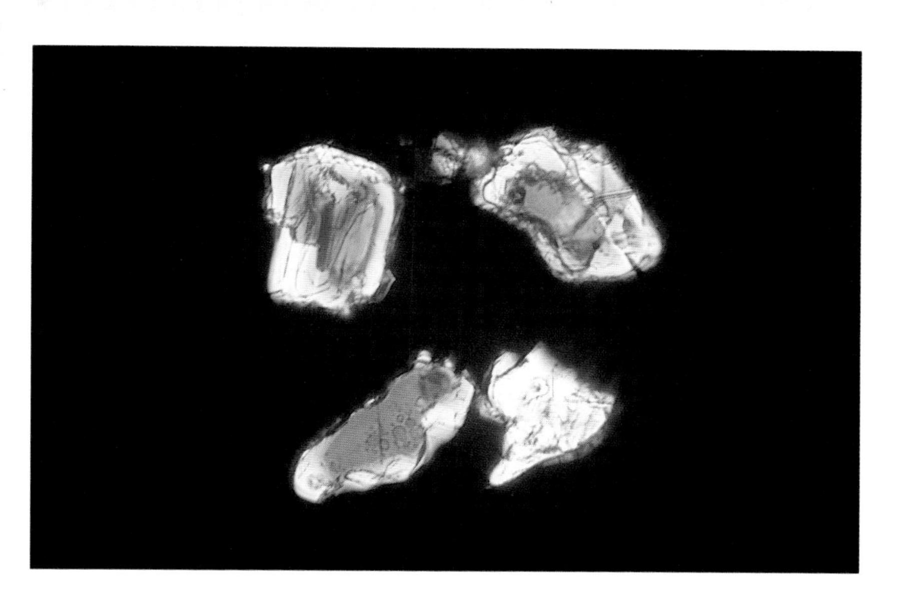

Anhydrite

$CaSO_4$
orthorhombic, biaxial (+)

$n\alpha$	1.569–1.574
$n\beta$	1.574–1.579
$n\gamma$	1.609–1.618
δ	0.04
Δ	2.90–3.00

Form in sediments: Grains are dominantly tabular cleavage fragments which often exhibit steplike margins or ragged edges. Spherical aggregates and fibrous radiating grains are less common. Anhydrite has mutually perpendicular cleavages in three directions. Twinning is common and multiple twinning in two directions, intersecting nearly at 90°, is visible on some grains. Fluid inclusions or organic impurities are not uncommon.

Colour: Grains are clear, transparent, colourless or, rarely, pink and pale orange.

Pleochroism: Non-pleochroic.

Birefringence: As a result of strong birefringence, anhydrite commonly exhibits 'twinkling' (a change of relief) as the microscope stage is rotated. Interference colours are vivid second- and third-order tints. They are often arranged in a 'patchy' fashion or are seen as several interference colour bands.

Extinction: Parallel to cleavage traces and to crystal elongation.

Interference figure: Cleavage fragments that lie on {100} give an acute bisectrix figure with several isochromatic curves; other orientations yield flash figures.

Elongation: Positive.

Distinguishing features: Pseudocubic cleavages, 'twinkling' and brilliant interference colours are distinctive. Baryte and celestite are likely to be mistaken for anhydrite. However, their relief is higher and 'twinkling' is absent. Opaque impurities are common in baryte and its extinction is symmetrical to prismatic cleavages. The habit of celestite is often irregular, 'sheaf' type or fibrous; it has only one perfect cleavage and often shows a wavy extinction. Brilliant interference colour bands are not seen on celestite or baryte.

Occurrence: Anhydrite is a characteristic constituent of evaporite sequences and it occurs either as a primary mineral or is formed by the dehydration of gypsum. In sedimentary rocks it is associated with dolomites and limestones, but is also present in sandstones, marls and shales. In metalliferous veins the oxidation of sulphides may form anhydrite.

Grains from: Rotliegend, northern Germany (Mmt 1.539).

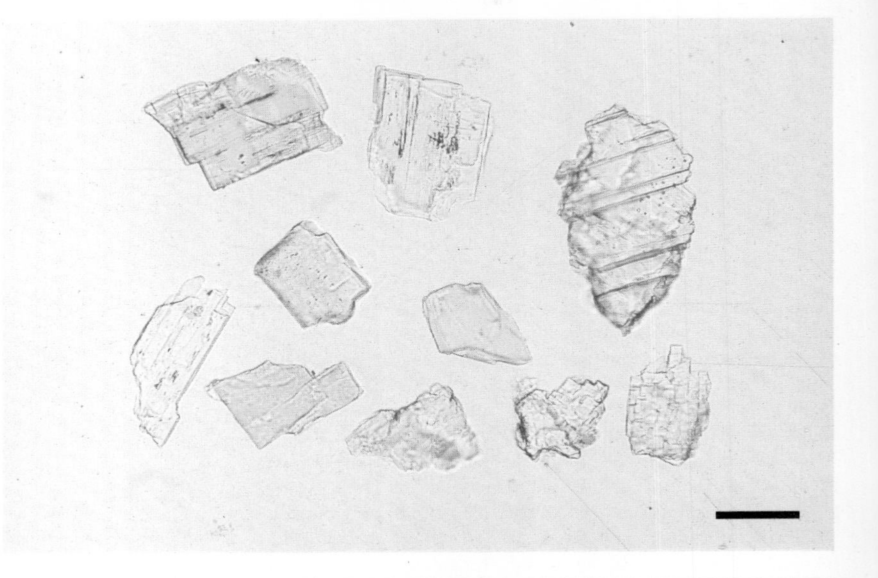

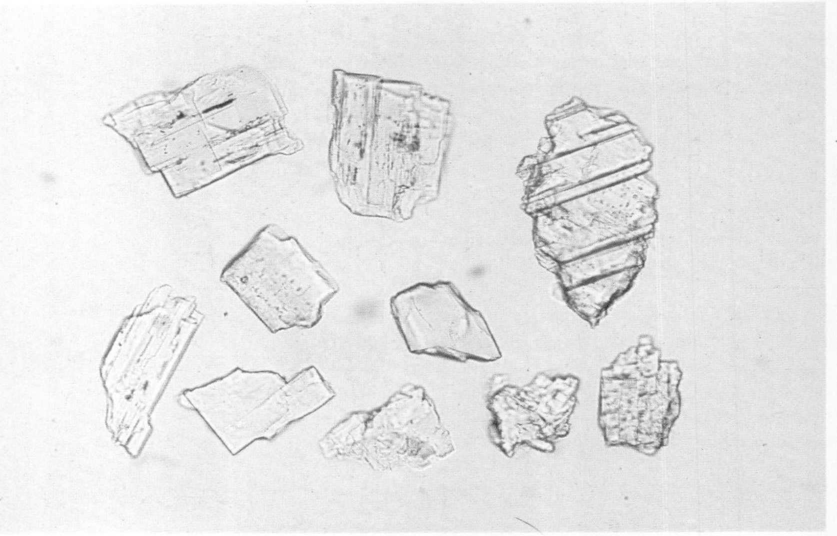

Jarosite

$K,Fe_3(SO_4)_2(OH)_6$
trigonal, uniaxial (−)

$n\omega$ 1.815–1.820
$n\varepsilon$ 1.713–1.715
δ 0.101–0.105
Δ 3.26

Form in sediments: Jarosite usually forms *'in situ'* and appears as euhedral crystals with a hexagonal habit. The edges of some grains may be curved, smoothed, or are surrounded by small platelets of subsequent growths and sometimes by scaly or fibrous aggregates. It also occurs as fragments of euhedral crystals, fine aggregates and clusters of fibres. Because of distinct basal cleavage, grains appear flaky, showing several superimposed plates. Zoning is very common. Sector twinning, composed of six segments, is often detected on basal flakes. Opaque impurities or limonite (as a decomposition product) are common.

Colour: Grains are golden yellow or honey yellow. Colour zoning is visible in many grains and it is manifested in a colourless core and a yellow outer rim.

Pleochroism: Jarosite prisms have been reported strongly pleochroic, but basal plates and aggregates lack pleochroism.

Birefringence: Strong, and prismatic fragments, aggregates and fibres show vivid high-order polarization colours. Basal flakes exhibit low-order interference tints which are often masked by strong mineral colour.

Extinction: Parallel to crystal faces and to fibres. Most basal sections show wavy extinction due to twinning or fail to extinguish completely.

Interference figure: Basal sections provide well-centred uniaxial figures in a yellowish-green field with a few isochromes or none. Occasionally, biaxial negative figures are obtained.

Elongation: Positive.

Distinguishing features: Jarosite is diagnosed by hexagonal form, platy basal habit, distinct basal cleavage, frequent sector twinning and by uniaxial character. Fibrous grains or aggregates and prismatic forms are usually found together with euhedral grains, thus ensuring identification. Owing to the flaky habit, jarosite resembles micas, but the distinctly higher relief, colour and twinning suffice to distinguish it from the latter.

Occurrence: Jarosite is a secondary mineral and is formed by the oxidation of iron-bearing sulphide ores. When encountered in sandstones and shales it is usually the alteration product of glauconite, pyrite or markasite.

Grains from: Triassic, northern Tunisia (CB).

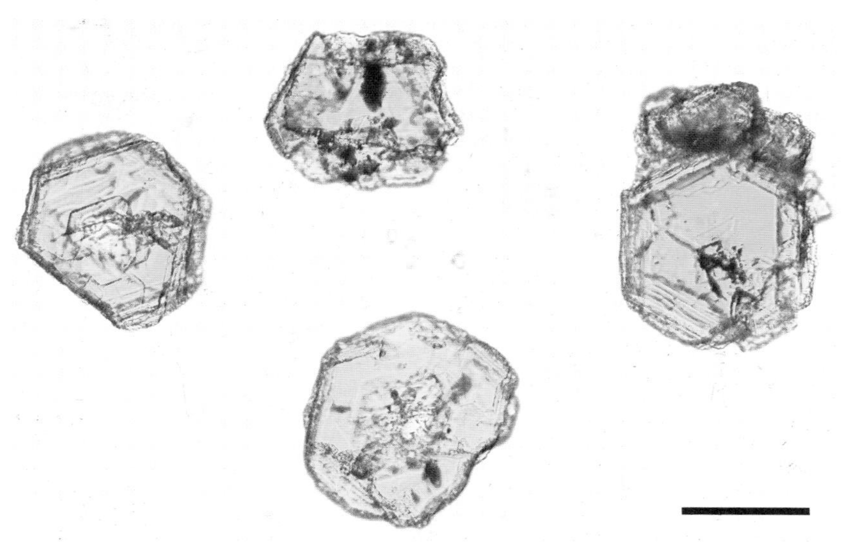

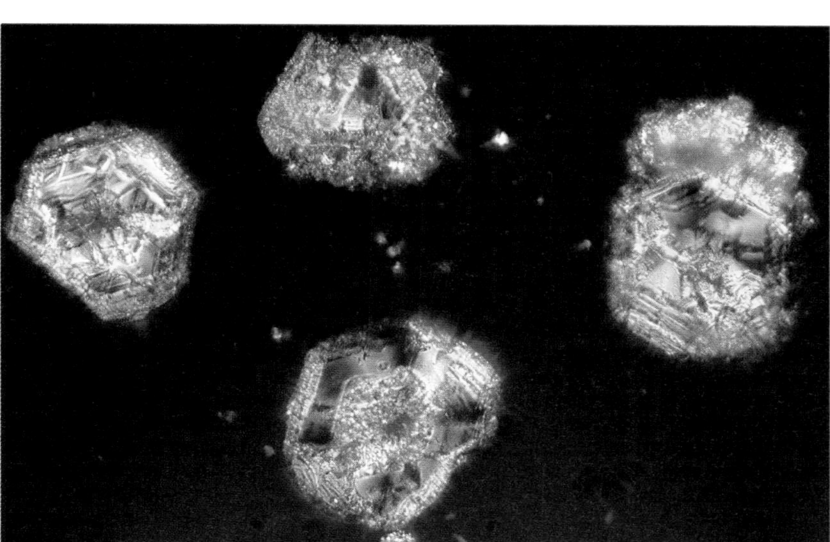

CARBONATES

Calcite

$CaCo_3$
trigonal, uniaxial (−)

$n\omega$ 1.486–1.550
$n\varepsilon$ 1.658–1.740
δ 0.172–0.190
Δ 2.715–2.94

Calcite is found in both the heavy and the light mineral fraction. The present description deals only with well crystallized calcite; ooliths, biogenic and compound detrital limestone debris are excluded.

Form in sediments: Calcite grains are generally irregular-angular or rectangular cleavage fragments. Skeletal and rounded grains or aggregates are also common, but rhombohedrons are infrequent. Perfect rhombohedral cleavage is sometimes visible. Lamellar twinning is characteristic of calcite, but intersecting twin lamellae, seen in thin sections, are seldom in detrital fragments. The twin-planes usually show blue or pink tints when the microscope stage is rotated.

Colour: Colourless, but finely dispersed impurities often cause a turbid appearance.

Pleochroism: Non-pleochroic.

Birefringence: Extremely high. When the microscope stage is rotated the relief shows a pronounced change ('twinkling'). In rhombic cleavage fragments, high relief appears when the long diagonal is parallel to the vibration direction of the polarizer. Interference colours are high-order pastel tints and the thinner edges of the grains may show numerous interference-colour bands. Thicker grains appear almost white under crossed polars. Those which fail to extinguish in any position show abnormal bluish-yellow interference colours.

Extinction: Symmetrical or inclined to cleavage traces or twin lamellae.

Interference figure: Detrital calcites usually yield an off-centre uniaxial interference figure with dense, thin isochromatic rings. Grains that fail to extinguish and show abnormal polarization colours have a biaxial figure.

Elongation: Rhombohedral fragments are length slow in the direction of the shorter rhomb diagonal, but because of extreme birefringence the sign of elongation is rarely discernible.

Distinguishing features: Extreme birefringence, 'twinkling' and very high-order interference colours are diagnostic. In grain mounts it is almost impossible to distinguish calcite from dolomite–ankerite, and complementary techniques (such as staining, X-ray or microprobe analyses) are necessary to enable diagnosis. The habit of dolomite is more commonly rhombohedral, but twinning in calcite is more common than in dolomite. Calcite is soluble in acetic acid and in dilute cold HCl; therefore, carbonate minerals remaining in the heavy residue after acid treatment are more likely to be either dolomite–ankerite or siderite.

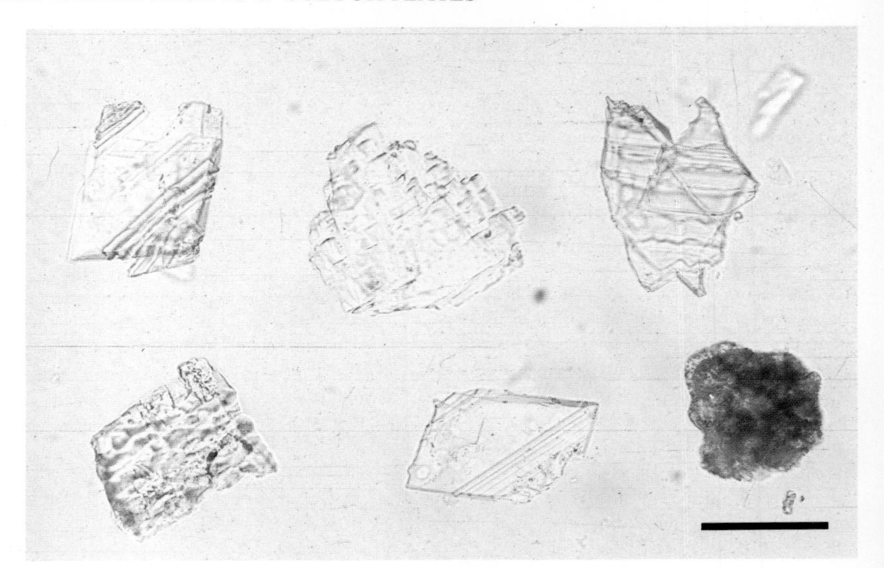

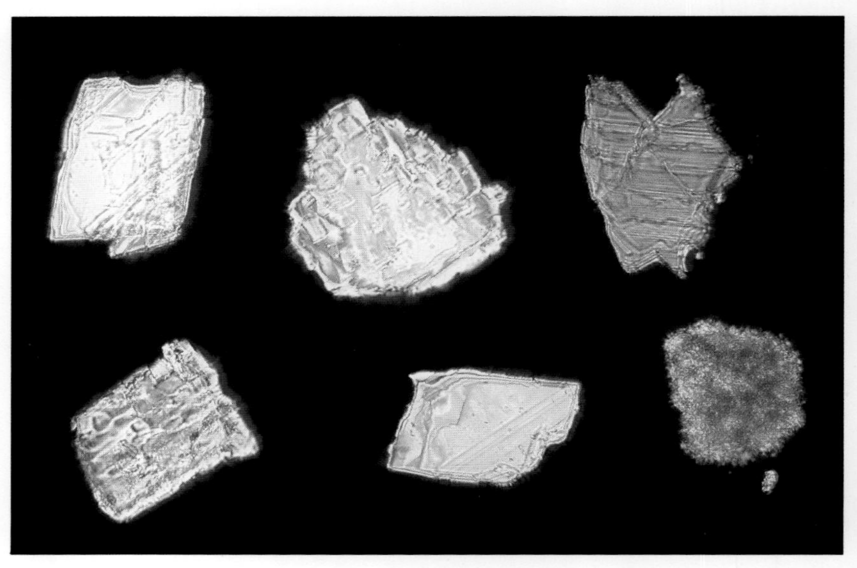

Occurrence: Details on the occurrence of calcite in sedimentary rocks is beyond the scope of this work. Calcite present in sandstones, greywackes, etc., is usually the cementing agent, a fissure filling or a replacement of other minerals. In igneous rocks calcite may occur as primary magmatic calcite. It is present in hydrothermal deposits and is a secondary mineral in cavities or amygdales of mafic volcanics. Calcite is a common alteration product of plagioclase. In metamorphic rocks it may be the principal constituent of marble and occurs in calc-schists and gneisses.

Grains from: Oligocene, Barrême Basin, France (Mmt 1.582).

Siderite

$FeCO_3$
trigonal, uniaxial (−)

$n\omega$	1.782–1.875
$n\varepsilon$	1.575–1.635
δ	0.207–0.242
Δ	3.50–3.96

Form in sediments: Siderite appears most commonly as rhombohedrons and sometimes as angular or rounded cleavage fragments. Nodular, botryoidal and oolithic forms are also frequent, as are aggregates. Lamellar twinning, though not common, may be present. As all trigonal carbonates, siderite has a perfect rhombohedral cleavage. As a result of oxidation the surface of the grains and cleavage cracks are often spotted or stained with a brownish-black substance.

Colour: Colourless or yellowish and brownish yellow.

Pleochroism: Non-pleochroic.

Birefringence: Extreme birefringence is characteristic. The relief changes on stage rotation ('twinkling'), similarly to calcite and dolomite, but in low-index resins (e.g. Canada balsam) this is not so well pronounced because its lower RI is higher than that of the immersing medium. Interference colours are white or pearl-grey, but may be masked by the common brownish stain. Transparent thin fragments display high-order shades with dense bright interference colour bands on the edges.

Extinction: Symmetrical to cleavage traces and rhombohedral faces, and parallel to the diagonals of the rhombohedron. Most botryoidal or oolithic siderites fail to extinguish.

Interference figure: Rhombohedral grains yield centred or slightly off-centre figures, usually with many fine isochromatic rings and broadening isogyres towards the periphery of the figure.

Distinguishing features: Rhombohedral or spherulitic form, all refractive indices greater than that of the Canada balsam, and (usually) brown stain – all serve to distinguish siderite from calcite and dolomite. It is relatively insoluble in acetic acid and in cold dilute HCl.

Occurrence: Siderite is associated with bedded sedimentary deposits, clay ironstones, banded cherty iron-carbonate rocks, siderite mudstones, sandstones and sideritic limestones. It also forms hydrothermally in metallic veins. Fissures and cavities in basalt, andesite and diabase may contain siderite.

Grains from: (a) River sand, Isère, France; (b) Rotliegend, The Netherlands; (c) Oligocene, Barrême Basin, France (Mmt 1.582).

Dolomite–ankerite series

$CaMg(CO_3)_2$ $Ca(Mg,Fe)(CO_3)_2$
trigonal, uniaxial (−)

	dolomite	ankerite
$n\omega$	1.679	1.690–1.750
$n\varepsilon$	1.500	1.510–1.548
δ	0.179	0.202
Δ	2.86	3.10

There is a continuous replacement of Mg by Fe which leads from dolomite through ferroan dolomite to ankerite. In grain mounts, members of this series are indistinguishable from each other.

Form in sediments: Rhombohedral and tabular morphologies are the most common. Grains occasionally have rounded corners. Dolomites with a hexagonal habit or irregular forms are rarely encountered. Crystals sometimes enclose organic matter, which may be zonally arranged. Rhombohedral cleavages are characteristic and cleavage traces often intersect in oblique angles. Twinned dolomite grains are infrequent.

Colour: Colourless, grey or turbid.

Pleochroism: Non-pleochroic.

Birefringence: Extreme, and grains 'twinkle' when the microscope stage is rotated. High relief appears when the long diagonal parallels the vibration direction of the polarizer. Interference colours are pearl grey or white. Bright fourth- or fifth-order pastel tints may be noticeable on thin edges. Basal sections usually show no 'twinkling' and display very bright yellow interference colours with a pale blue hue. These grains fail to extinguish in any position.

Extinction: Symmetrical to the rhombohedron faces and parallel to its diagonals.

Interference figure: Rhombohedral crystals yield off-centre interference figures. Basal sections have well-centred uniaxial figures with many isochromatic rings. The isogyres show a characteristic broadening towards the periphery of the figure. Owing to strain, dolomites may have small 2V.

Elongation: Elongated grains usually have negative elongation, but the high-order interference colours make observation difficult.

Distinguishing features: Pronounced change of relief, extreme birefringence and rhombohedral form are characteristic, but only complementary techniques can prove a positive identification of members of the dolomite–ankerite series, or distinguish these from other carbonate minerals. As calcite is soluble in acetic acid and in dilute cold HCl, residual carbonate grains (after acid treatment) are likely to be dolomite–ankerite or siderite. Rhombohedral or hexagonal habit is more typical of dolomite than of calcite, and twinning is more frequent in the latter. Siderite has a slightly stronger relief, frequent brown staining, and its lower RI is higher than that of the Canada balsam. Rounded grains with anomalous interference colours may resemble sphene, but the latter is biaxial.

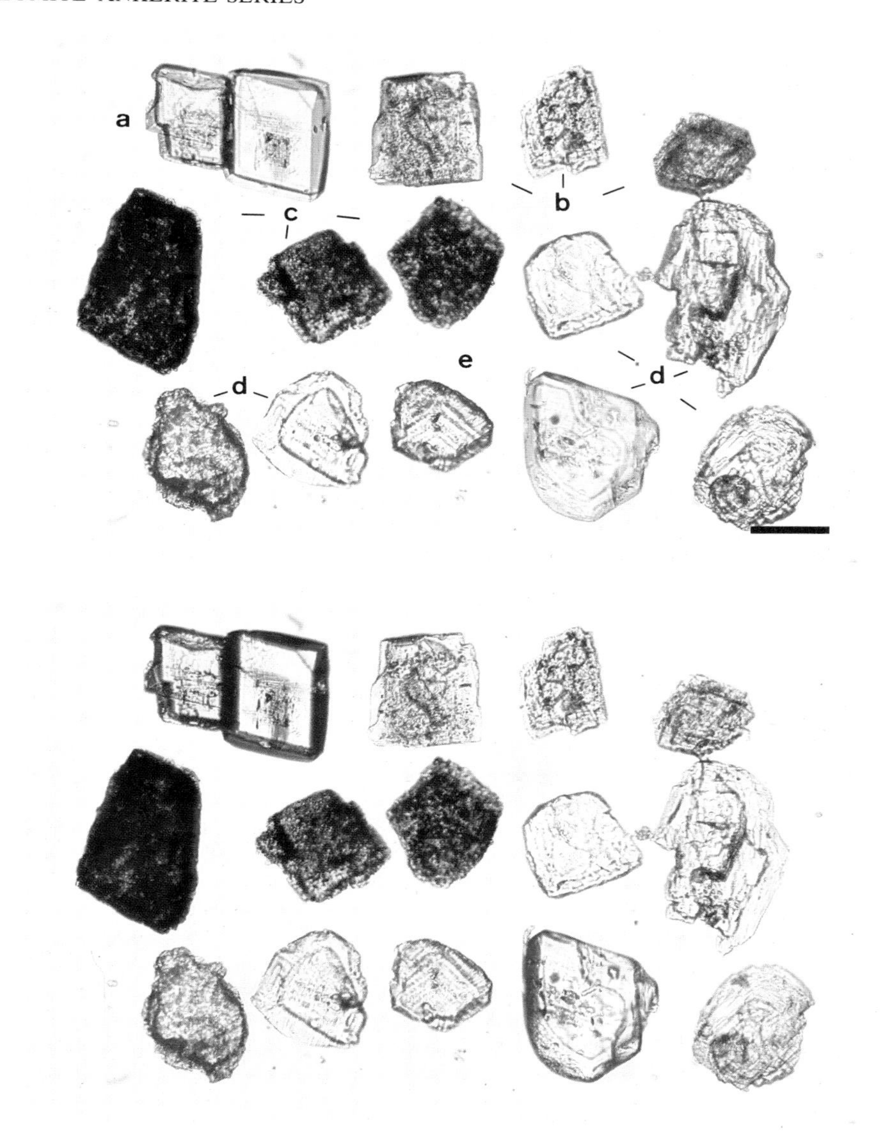

Occurrence: Dolomite is widespread in almost all types of sedimentary rocks. Contact or regional metamorphism produces dolomitic marble from dolomites or magnesian limestones. It is found in altered ultrabasic rocks, in hydrothermal veins, in carbonatites and in highly alkaline dykes. The occurrence of ankerite is more restricted and it usually appears as veins and concretions, and occasionally as cement in Fe-rich sediments.

Grains from: (a) Dolomite, Rotliegend, northern Germany; (b) ankerite, Jurassic, North Sea; (c) ankerite, Buntsandstein, Triassic, borehole Pfaffnau, 1823 m, Switzerland; (d) dolomite, river sand Isère, France; (e) dolomite, Oligocene Molasse, Savoy, France (Mmt 1.582).

PHOSPHATES

Apatite minerals

hexagonal, uniaxial (−)

fluorapatite	$Ca_5(PO_4)_3F$
hydroxyapatite	$Ca_5(PO_4)_3OH$
chlorapatite	$Ca_5(PO_4)_3Cl$
carbonate–apatite	$Ca_5(PO_4CO_3OH)_3(F,OH)$?

$n\omega$	$n\varepsilon$
1.633–1.650	1.629–1.646
1.643–1.658	1.637–1.654
1.650–1.667	1.647–1.665
1.618	1.631

δ 0.001–0.007

Δ 3.10–3.35

Form in sediments: Apatite usually concentrates in the finer grades. Euhedral crystals are short stubby or long slender prisms with either simple pyramidal terminations or a combination of the basal pinacoid and pyramids (upper). Occasionally well-developed hexagonal basal plates are encountered (second page, upper, a). It also occurs as broken prisms and rectangular or sometimes irregular fragments. However, grains most commonly exhibit rounding which may be incipient, seen as smooth or curved terminations or edges, but the most frequent are the extensively rounded, egg-shaped or spherical morphologies (second page, upper, d and middle). Cleavages are poor. Inclusions are magnetite, titanite and zircon, as well as dense minute grey-to-opaque needles which are orientated parallel to the *c* axis. Strings of fluid globules and striations are also common. Apatite grains may enclose a 'smoky' dark pleochroic core (second page, upper, c). Overgrowth on detrital apatites are rarely reported, presumably because they are not as readily discernible as those of zircons and tourmalines. The overgrowth are either small sharp triangular projections attached to certain parts of the grain or they enclose the detrital grain completely, evolving into an euhedral crystal. According to our observations the overgrowth tends to form around the basal pinacoid. When it is fully developed, the grain attains a hexagonal habit, showing the combinations of the basal pinacoid and short pyramids. Within this form the rounded and pitted, sometimes iron-oxide pigmented, detrital grain nucleus is clearly visible. The overgrowth is in optical continuity with the host grain but its RI may be higher, indicating a differing chemistry. Authigenically formed apatites are generally hexagonal plates. They may be zoned concentrically and divided into six sectors (second page, lower, a). Apatites are frequently corroded, which is manifested either by dense surface etch pits, mamillae and ragged edges, or by hollows and grooves entering deep into the grain (second page, upper, e and f).

Colour: Grains are usually colourless, but because of the presence of manganese, ferrous or ferric irons, or rare earths, some apatites may have greyish-greenish or reddish-brown tints (upper, b).

Pleochroism: Reddish-coloured varieties are distinctly pleochroic in shades of: ω, pale reddish or yellowish brown; ε, dark red-brown. The 'dusty' cores of some apatites are usually pleochroic.

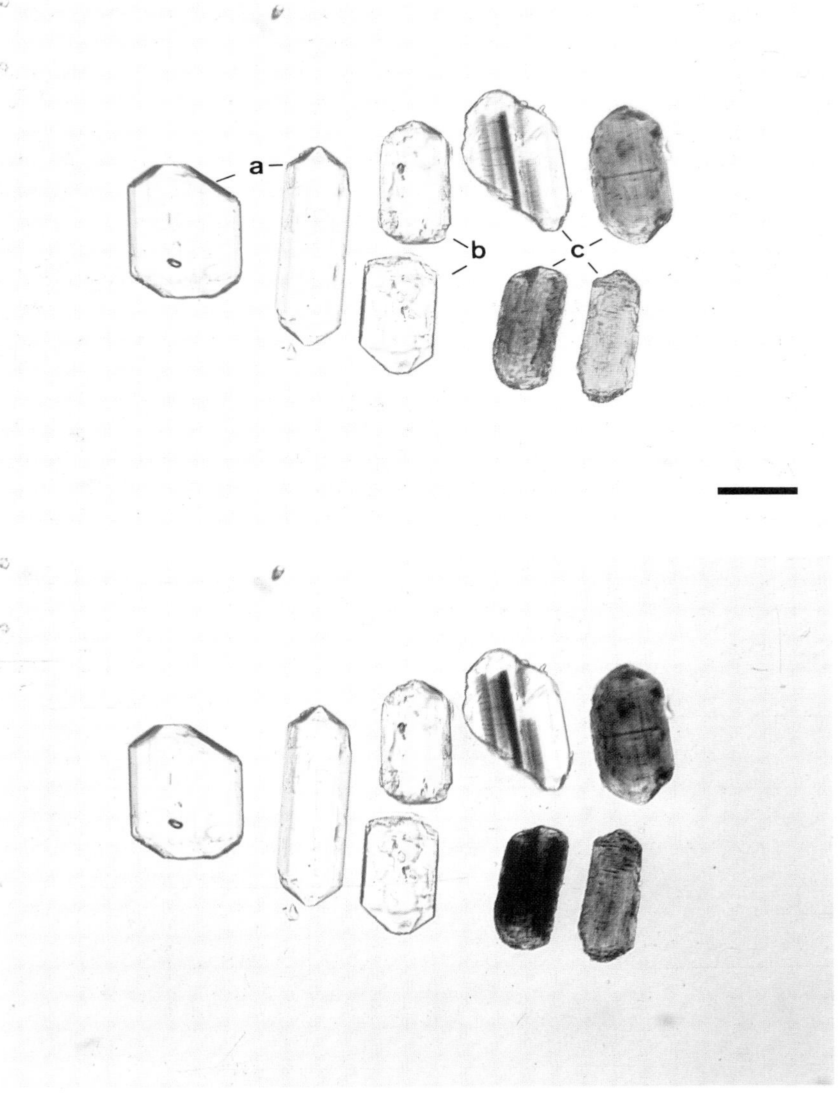

Birefringence: Maximum birefringence is weak to moderate. Interference colours are first-order dark grey or pale whitish grey. Very thick grains display first-order yellow or orange colours and their thicker centre part may be blue or mauve.

Extinction: Prismatic crystals and rounded oblong grains have parallel extinction.

Interference figure: Most apatite grains show anomalous biaxial interference figures with a positive sign. Hexagonal basal sections and non-birefringent grains (orientated according to the basal pinacoid) yield well centred uniaxial figures in a dark-grey field with a negative sign.

Elongation: Negative.

Distinguishing features: Moderately high relief, general lack of colour, characteristic morphology and weak birefringence are distinctive. Quartz has a considerably lower relief. The good basal cleavage, lower relief and usually brighter interference colours of topaz distinguish it from apatite. Andalusite is pleochroic in pink shades, has a lower RI, an irregular morphology and frequently encloses carbonaceous impurities. The birefringence of sillimanite is stronger and is length slow. Zoisite has higher relief, strong dispersion, positive elongation and often displays anomalous interference colours. Colourless tourmaline has a markedly stronger birefringence. Amongst the apatite minerals the highest relief of chlorapatite may aid distinction.

The distinction of apatite grains from baryte in baryte-contaminated drill-cutting samples is described under baryte.

Carbonate fluorapatite is called **francolite** (upper, g) and it occurs as brown- or yellowish-coloured spherules, fibres or fossil replacements; it shows low-order interference tints as well as undulatory extinction. **Collophane** (upper, h) is brown and oolitic, globular or botyroidal. It is cryptocrystalline and isotropic.

Occurrence: Apatite is a common accessory mineral of almost all types of igneous rocks. Amongst the apatite minerals, fluorapatite is the most abundant and is usually related to granitic rocks. Chlorapatite is more common in the basic types. Apatites form in carbonatites, hydrothermal veins and cavities, as well as in both thermally and regionally metamorphosed rocks. Apatite may form authigenically (Weissbrod & Nachmias 1986, and grains shown in lower, and on third page).

Apatite minerals are widespread in sediments. They occur dominantly as detrital grains, but may be the cementing medium of sandstones, or they form potentially economic phosphate deposits.

Remarks: Apatite dissolves in acids; therefore, during sample preparation strong acid treatment should be avoided (p. 11). Under deep burial (Table 2.1) apatite is stable; it may survive several cycles of reworking and is common in ancient sediments. Because it is unstable in acids, acidic geochemical environments are often depleted in apatites (Morton 1984b, 1986).

Apatite is used in fission-track dating. It also conveys important clues to the timing of oil generation in hydrocarbon source rocks (Gleadow *et al.* 1983).

Grains from: First page, upper: (a) Upper Eocene tuff

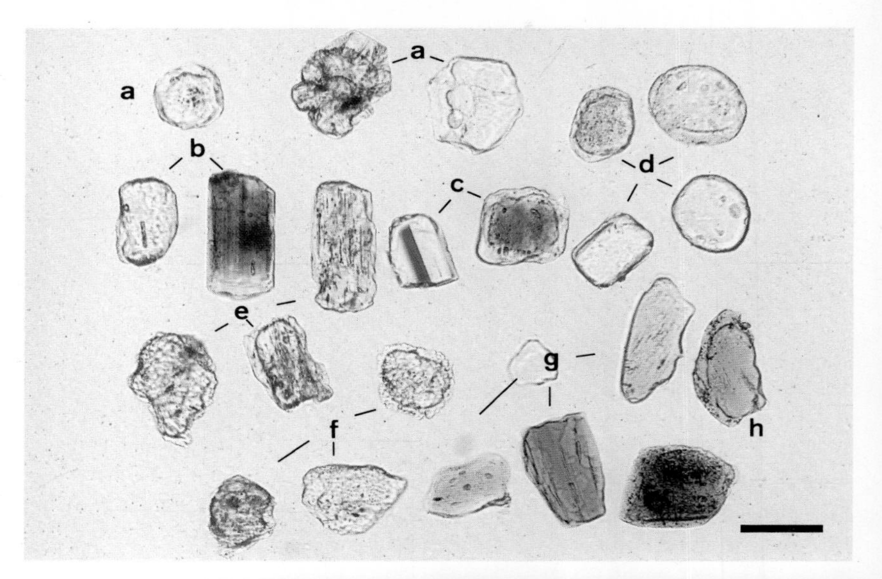

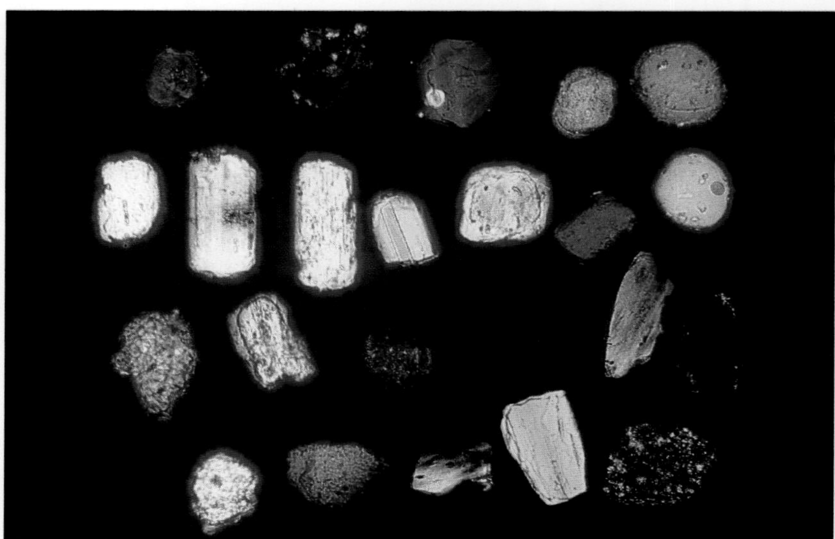

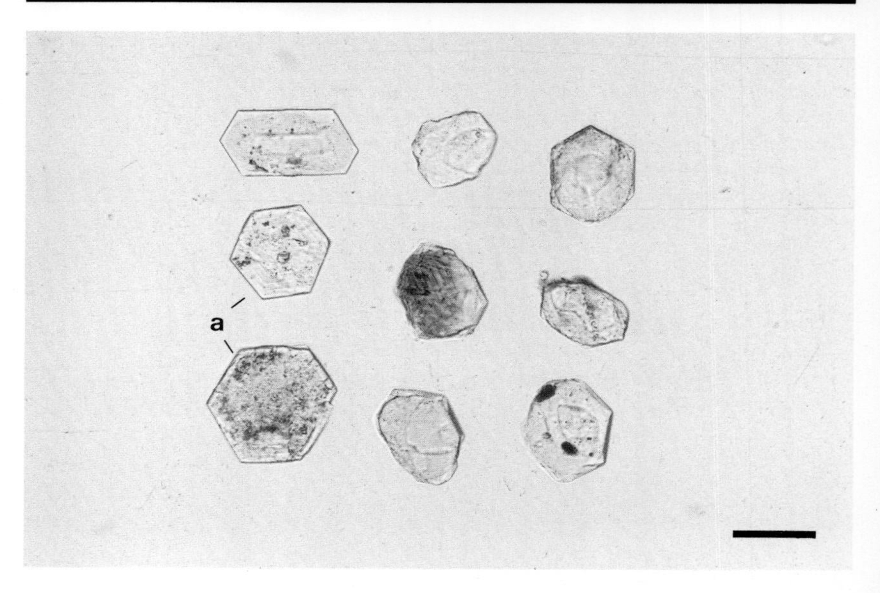

horizon, Possagno, Italy; (b) Oligocene, Barrême Basin, France.
Second page, upper: (a) River sand, Isère, France; (b) Jurassic, North Sea; (c) Oligocene Molasse, borehole Romanens-1, 1890 m, Switzerland; (d) Rotliegend, northern Germany; (e) Cretaceous, Tunisia; (f) Buntsandstein, Triassic, borehole Riniken, 811 m, Switzerland; (g) francolite and (h) collophane, Oligocene, Nerthe-chain, southern France.
Second page, lower, a: Authigenic grains, Buntsandstein, Triassic, borehole Kreuzlingen, 2532 m, Switzerland; Other grains with overgrowth: Rotliegend, northern Germany.
All grains are embedded in Mmt 1.582.

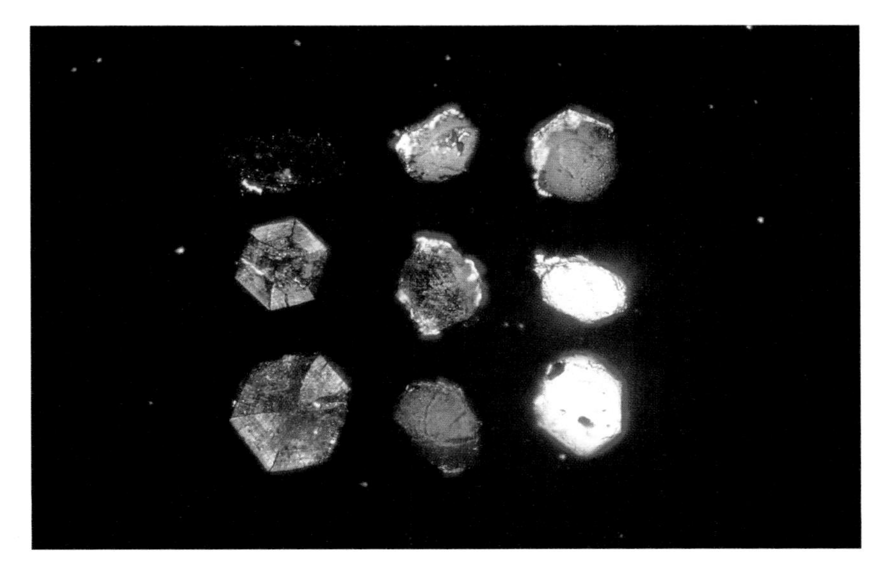

Monazite

(Ce,La,Th)PO_4
monoclinic, biaxial (+)

$n\alpha$ 1.774–1.800
$n\beta$ 1.777–1.801
$n\gamma$ 1.828–1.851
δ 0.045–0.075
Δ 5.0–5.3

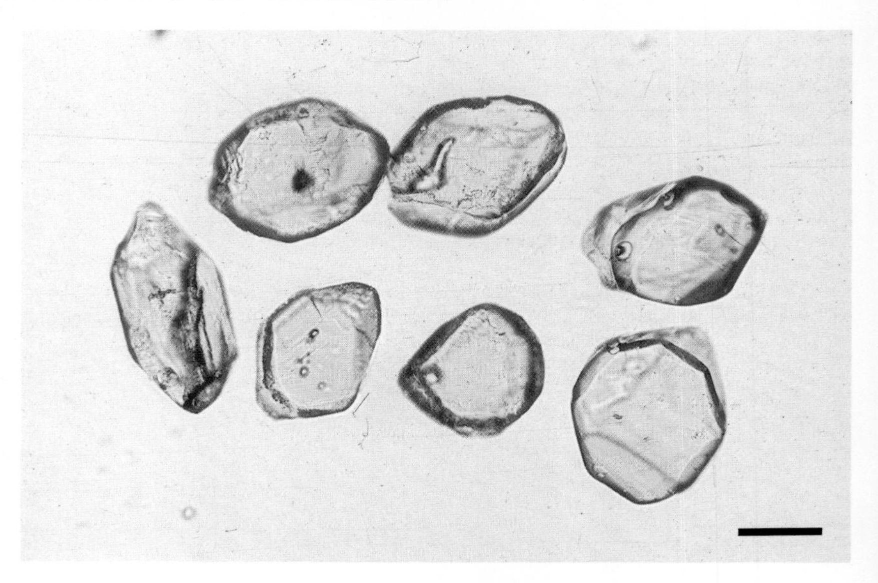

Form in sediments: Monazite has a high relief, resinous lustre and it appears with a dark rim surrounding its outline. Grains are dominantly well rounded, egg-shaped or spherical. Subrounded or euhedral grains have been reported from alluvial concentrates (upper). Dense surface pitting or a yellowish-brown stain are often detectable. Some grains may enclose zircon and rutile inclusions, opaque substances and fluid globules.

Colour: Pale yellow, almost colourless, pale amber, greenish yellow and rarely brown.

Pleochroism: Deep-coloured thicker grains may display weak pleochroism in shades of yellow or, rarely, green.

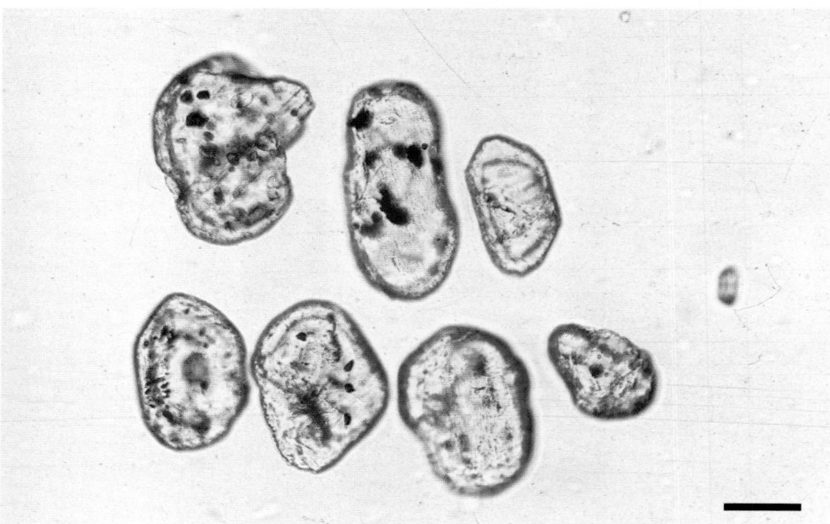

Birefringence: Strong, and interference colours range to upper third- or fourth-order pearl white or yellow. The well-rounded grains show distinct interference colour bands.

Extinction: Maximum extinction is 2°–7°, but detrital grains usually fail to show a complete extinction.

Interference figure: Grains lying on the basal parting plane (recognized by weaker interference colours and incomplete extinction) yield excellent well-centred or nearly centred acute bisectrix figures with small 2V. The well-defined isogyres appear in a yellow or greyish-white field. Isochromes are either few or numerous. Dispersion is weak. Cleavage fragments yield flash figures.

Elongation: {100} faces elongate parallel with *c* have positive, and crystal fragments elongated on *b* have negative, elongation.

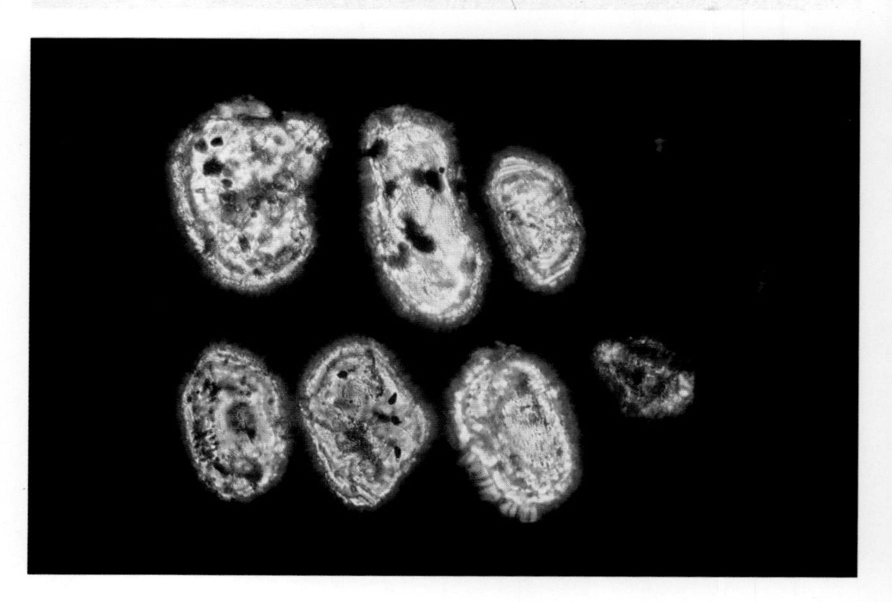

Distinguishing features: Monazite is diagnosed by high relief, colour and mostly rounded morphology. The frequent brownish stain and surface pitting are also characteristic. The only detrital minerals for which monazite is likely to be mistaken are yellow zircon, xenotime and, more rarely, sphene. The relief of zircon is higher, and surface etching is uncommon in zircon. Sphene has more intense polarization colours, strong dispersion and higher refractive indices. Xenotime, with which monazite is often associated in sediments, strongly resembles monazite. The former displays higher-order interference tints and is uniaxial. A positive identification of monazite and its distinction from xenotime can be made by means of a 'pocket spectroscope' (Adams 1954, Smithson 1959, Hering & Zimmerle 1963). Strong broad double, less commonly triple, absorption lines in the yellow (caused by neodymium and amplified by praseodymium), a fairly strong band in the green, and less commonly in the blue-green, are diagnostic of monazite (see also under xenotime).

Occurrence: Monazite is an accessory mineral of granitic

rocks. Larger crystals occur in granitic and syenitic pegmatites and in Alpine-type veins. It is a rare constituent of metamorphic schists, gneisses and granulites. Monazite is fairly resistant to weathering and diagenetic conditions. Concentrates of monazite are known from fluvial and beach sands, and it also occurs in tin placers.

Grains from: Upper: River sand, Centovalli, Switzerland (Mmt 1.662); Middle: Buntsandstein, Triassic, Mumpf, Switzerland (Mmt 1.662).

Xenotime

YPO_4
tetragonal, uniaxial (+)

$n\omega$	1.719–1.724
$n\varepsilon$	1.816–1.827
δ	0.097–0.103
Δ	4.59

Form in sediments: Detrital species occur as doubly terminated stumpy euhedral crystals with a simple habit (b), short prisms, rectangular or fractured irregular fragments, basal forms (c) and rounded to well-rounded grains. They exhibit a high relief. Inclusions are zircon and opaque impurities. Some grains may display zoning. Unaltered xenotime appears with a high resinous or vitreous lustre, but grains may be densely pitted or speckled with brownish alteration products which also appear in anastomosing cracks and in irregularities.

Colour: Colourless, honey yellow, pale brown or very pale green.

Pleochroism: Weak pleochroism is discernible in deeper-coloured varieties: ω, pale yellow, pale yellow brown, pale rose; ε, pale yellowish green, yellow, grey brown.

Birefringence: Very strong and 'twinkling' may appear on stage rotation. Interference colours are high-order pearl white or pink, similar to those of calcite. Intensive mineral colour can mask the interference tints. Thinner edges sometimes display bright narrow interference-colour bands.

Extinction: Parallel.

Interference figure: Prismatic faces give faint biaxial flash figures. Basal forms (such as shown in c) yield excellent centred uniaxial figures with several dense narrow isochromes. These grains usually fail to extinguish and display bluish-yellow polarization colours. Near-basal sections and some rounded grains provide off-centre uniaxial or biaxial figures with small 2V.

Elongation: Positive.

Distinguishing features: The similarity of xenotime to zircon, monazite and sometimes to sphene, causes some difficulties in diagnosis. When only optical methods are applied, the simple habit of the euhedral grains compared to that of euhedral zircon, and xenotime's lower refractive indices, higher birefringence and, when present, brownish alteration products, aid distinction. Xenotime strongly resembles monazite, and positive identification can be made only when a centred interference figure is obtainable, as monazite is biaxial and xenotime is uniaxial. The 2V of monazite is small and its off-centre figures may resemble the off-centre figures of uniaxial minerals, enhancing confusion. However, the very high-order interference colours of xenotime help diagnosis. Sphene has a stronger relief and dispersion and is biaxial.

Of the complementary methods used for identification, Hutton (1947) suggested various immersion media, but the simple, quick and most reliable approach for the identification of xenotime, and for distinguishing between xenotime, mona-

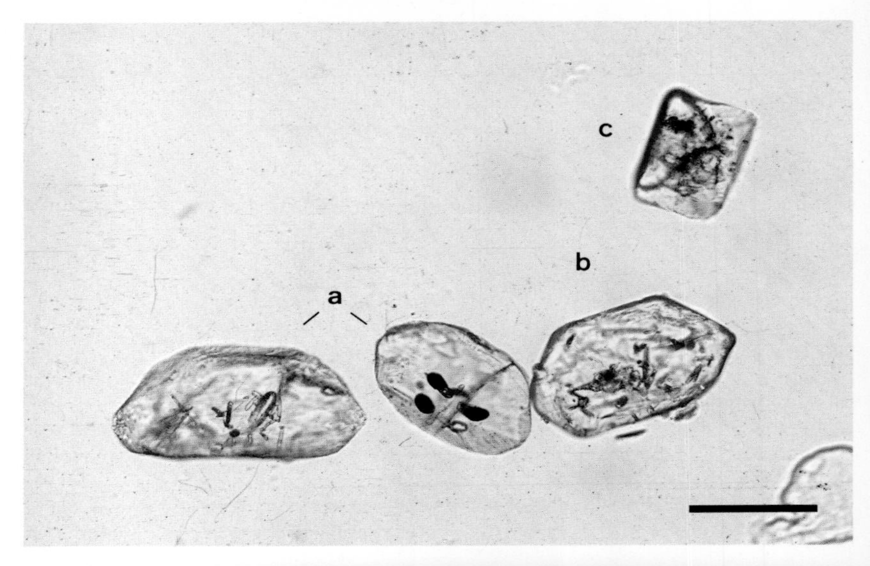

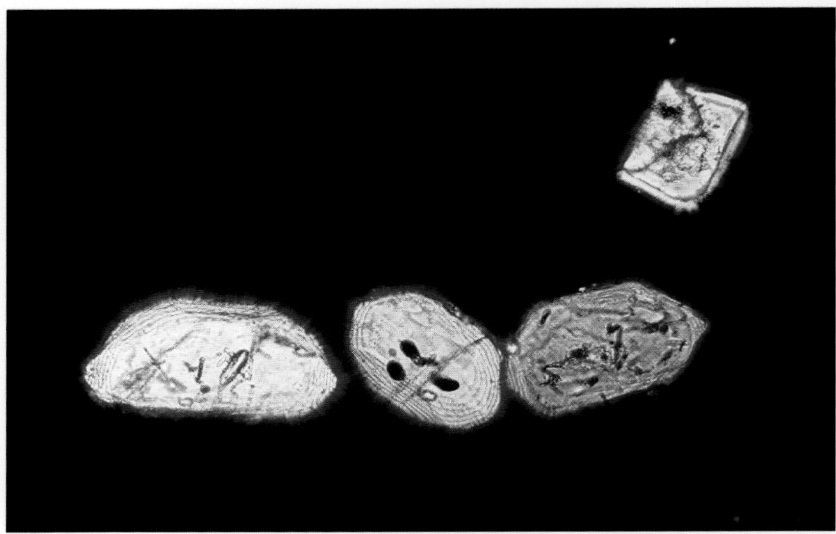

zite and zircon, is to use a simple 'pocket spectroscope' (Adams 1954, Smithson 1959). This, by revealing the distinct absorption spectra of xenotime and monazite, ensures a positive diagnosis. Hering & Zimmerle (1963) called attention to this method, discussed its validity, and gave data of the absorption lines of monazite and xenotime. Xenotime exhibits a strong thick band in the deep red and has two or three faint lines on both sides. A strong double line appears in the green, and in thicker grains a faint line is visible in the deep blue. It has no absorption lines in the yellow, in which monazite shows broad double, less commonly triple lines. Zircon fails to show any absorption lines. This method is strongly recommended.

Occurrence: Xenotime is an accessory mineral of granites, syenites and granite pegmatites. It is also a constituent of quartzose micaceous gneisses and of some Alpine-type veins. In placer deposits it concentrates with monazite, cassiterite, gold or diamond. Xenotime is fairly stable in sediments and is probably more widespread than reported in heavy mineral analyses. Monazite or zircon-rich assemblages may signal its occurrence.

Grains from: (a) Buntsandstein, Triassic, Mumpf, Switzerland; (b) Oligocene, Nerthe-chain, southern France; (c) Buntsandstein, borehole Pfaffnau, 1822 m, Switzerland (Mmt 1.662).

MISCELLANEOUS

Scheelite

$CaWO_4$
tetragonal, uniaxial (+)

$n\omega$ 1.918–1.920
$n\varepsilon$ 1.934–1.937
δ 0.016–0.017
Δ 5.94–6.12

Form in sediments: Grains have a high relief and are prismatic, irregular, subrounded or rounded. Generally they are thick and coarse grained. Etch-facets and multipyramidal terminations, similar to those exhibited by garnet, are frequent on detrital grains. Sometimes striae or cleavage traces are visible.

Colour: Colourless, grey, pale yellow or brownish. Most grains have a somewhat dark, 'dusky' appearance.

Pleochroism: Non-pleochroic.

Birefringence: Maximum birefringence is moderate and interference colours are grey or vivid yellow, blue and crimson, depending on the thickness of the grains.

Extinction: Elongated fragments have parallel extinction.

Interference figure: Grains usually yield an off-centre uniaxial figure with broad isochromatic curves and show a clear positive sign.

Elongation: Positive.

Distinguishing features: High relief, 'dusky' appearance, frequent etch-facets together with vivid interference colours are diagnostic. Cassiterite resembles scheelite, but it has a higher relief and birefringence. Garnet is isotropic and the anomalously anisotropic garnets have dull low-order interference colours and undulatory extinction.

Occurrence: Scheelite is formed in pegmatites and in high-temperature veins. It is present in contact-metamorphic rocks around granitic intrusives, and in amphibolites and regionally metamorphosed impure limestones.

Remarks: Pure scheelite shows a characteristic blue fluorescence in short wave UV-light. Mo-bearing scheelite has white to yellow fluorescence. It decomposes in HCl, leaving a yellow powdery residue of WO_3.

Grains from: River sand, Centovalli, Switzerland (Mmt 1.662).

Fluorite

CaF_2
isometric

n 1.433–1.435
Δ 3.18

Form in sediments: Owing to its low refractive index, fluorite appears with a negative relief even in low-index mounting media and its outline is surrounded by a pale yellow rim. Grains are triangular, rectangular or irregular fragments, usually with sharp (more rarely with curved) edges. Rounded grains are rare. Fluorite is relatively soluble in carbonated waters and the surface of the grains may be etched, showing triangular facets or crescent-shaped indentations. Cleavage traces are rarely noticeable. Iron ore and fluid inclusions may be present.

Colour: Colourless, pale violet or pale pink.

Distinguishing features: Negative relief together with isotropic character are diagnostic of fluorite and distinguish it from all other minerals.

Occurrence: Fluorite commonly occurs in hydrothermal, pegmatitic and pneumatolithic veins, in greisen, in cavities of granite, and occasionally in carbonate rocks and phosphorites. It is found sometimes in sandstones and rarely, it forms their cement.

Remarks: Fluorite from drilling mud can contaminate the heavy mineral fractions of drill cuttings.

Grains from: (a) Beach sand, St Ives, Cornwall, England; (b) Upper Eocene, Barrême Basin, France; (c) Buntsandstein, Triassic, borehole Böttstein, 314 m, Switzerland (Mmt 1.539).

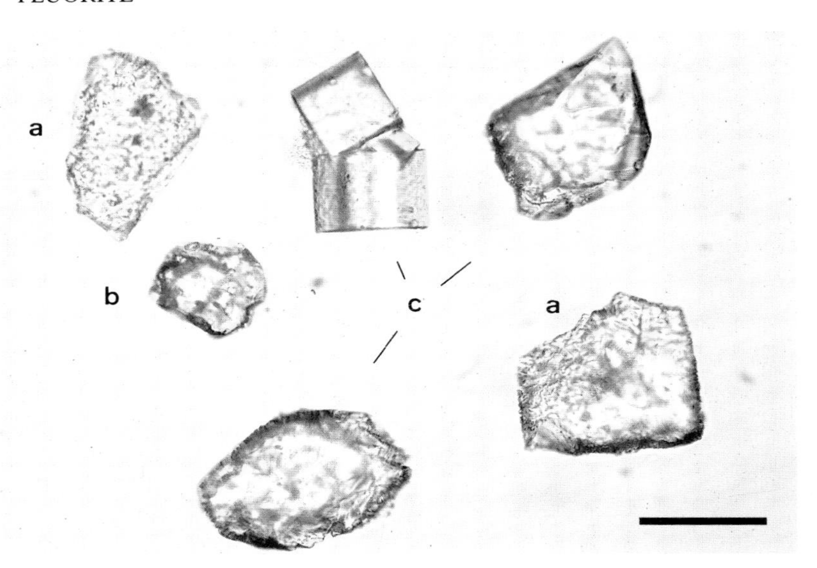

REFERENCES

Adams, J. W. 1954. A simple microspectroscope. *Am. Miner.* **39,** 393–4.

Alam, M. & B. J. W. Piper 1981. Detrital mineralogy and petrology of deep-water continental margin sediments off Newfoundland. *Can. J. Earth Sci.* **18,** 1336–46.

Aldahan, A. A. & S. Morad 1986. Authigenic sphene in sandstones of the Brottum Formation (Norway) and the Dala Sandstone (Sweden). *N. Jb. Miner. Mh.* **3,** 135–44.

Allen, P. A. & M. A. Mange-Rajetzky 1982. Sediment dispersal and palaeohydraulics of Oligocene rivers in the eastern Ebro basin. *Sedimentology* **29,** 705–16.

Allen, P. A. & M. A. Mange-Rajetzky 1989. Sedimentary evolution of a Devono-Carboniferous Rift Basin, Clair Field, UK: Impact of changing provenance. *Developments in Sedimentary Provenance Studies.* J. Geol. Soc. Lond. Spec. Publ., in preparation.

Allen, P. A., M. A. Mange-Rajetzky, A. Matter & P. Homewood 1985. Dynamic palaeogeography of the open Burdigalian seaway, Swiss Molasse basin. *Ecologae geol. Helv.* **78,** 351–81.

Allen, V. T. 1948. Weathering and heavy minerals. *J. Sedim. Petrol.* **18,** 38–42.

Allman, M. & D. F. Lawrence 1972. *Geological laboratory techniques.* London: Blandford.

Alty, S. W. 1933. Some properties of authigenic tourmaline from Lower Devonian sediments. *Am. Miner.* **43,** 351–5.

Anand, R. R. & R. J. Gilkes 1984. Weathering of hornblende, plagioclase and chlorite in meta-dolerite, Australia. *Geoderma* **34,** 261–80.

Anderton, R., P. H. Bridges, M. R. Leeder & B. W. Sellwood 1979. *A dynamic stratigraphy of the British Isles. A study in crustal evolution.* London: George Allen & Unwin.

Artini, E. 1898. Intorno alla composizione mineralogica delle sabbie di alcuni fiumi del Veneto, con applicazione ai terreni di transporto. *Riv. Miner. Crist. Italiana* **19,** 33–94.

Baak, T. A. 1936. *Regional petrology of the southern North Sea.* Wageningen: H. Veenman en Zonen.

Baker, G. 1956. Sand drift at Portland, Victoria. *R. Soc. Victoria Proc.* **68,** 151–97.

Baker, G. 1962. *Detrital heavy minerals in natural accumulates.* Australasian Inst. Mining Met. Proc., no. 1.

Barrie, J. V. 1980. Heavy mineral distribution in bottom sediments of the Bristol Channel, U.K. *Estuarine and Coastal Marine Sci.* **11,** 369–81.

Bekker, Yu., N. B. Bekasova & A. D. Ishov 1970. Diamond-bearing placers in the Devonian rocks of the Northern Urals. *Lithol. Miner. Resources* **4,** 421–30. (Translated from Litologiya i poleznye Iskopaemye **4,** 65–75.)

Belfiore, A. 1981. Heavy mineral dispersal in the Sardinia Basin (Thyrrhenian Sea). In *Sedimentary basins of Mediterranean margins,* F. C. Wezel (ed.), 271–82. Proc. C. N. R. Conference, Univ. Urbino, Pitagora.

Berner, R. A. 1978. Rate control of mineral dissolution under earth surface conditions. *Am. J. Sci.* **218,** 1235–52.

Berner, R. A. & J. Schott 1982. Mechanism of pyroxene and amphibole weathering. II. Observations of soil grains. *Am. J. Sci.* **282,** 1214–31.

Beveridge, A. J. 1960. Heavy minerals in Lower Tertiary formations in the Santa Cruz Mountains, California. *J. Sedim. Petrol.* **30,** 513–37.

Blatt, H. 1967. Provenance determinations and recycling of sediments. *J. Sedim. Petrol.* **37,** 1031–44.

Blatt, H. 1985. Provenance studies and mudrocks. *J. Sedim. Petrol.* **55,** 69–75.

Blatt, H. & B. Sutherland 1969. Intrastratal solution and non-opaque heavy minerals in shales. *J. Sedim. Petrol.* **39,** 591–600.

Bloss, F. D. 1981. *The spindle stage: principles and practice.* Cambridge: Cambridge University Press.

Blumenthal, W. B. 1958. *The chemical behaviour of zirconium.* New York: Van Nostrand.

Boenigk, K. 1983. *Schwermineralanalyse.* Stuttgart: Ferdinand Enke.

Borg, G. 1986. Facetted garnets formed by etching. Examples from sandstones of late Triassic age, South Germany. *Sedimentology* **33,** 141–6.

Boswell, P. G. H. 1933. *On the mineralogy of sedimentary rocks.* London: Thomas Murby.

Bramlette, M. N. 1929. Natural etching of detrital garnet. *Am. Miner.* **14,** 336–7.

Bramlette, M. N. 1941. The stability of minerals in sandstone. *J. Sedim. Petrol.* **11,** 32–6.

Brammal, A. 1928. Dartmoor detritals: a study in provenance. *Proc. Geol. Ass.* **39,** 27–48.

Briggs, L. I. 1965. Heavy mineral correlations and provenances. *J. Sedim. Petrol.* **35,** 939–55.

Briggs, L. I., D. S. McCulloch & F. Moser 1962. The hydraulic shape of sand particles. *J. Sedim. Petrol.* **32,** 645–56.

Brix, M. R. 1981. *Schwermineralanalyse und andere sedimentologische Untersuchungen als Beitrag zur Rekonstruktion der strukturellen Entwicklung des westlichen Hohen Atlas (Marokko).* Unpublished PhD thesis, University of Bonn.

Brückner, W. D. & H. J. Morgan 1964. Heavy mineral distribution on the continental shelf off Accra, Ghana, West Africa. In *Deltaic and shallow marine deposits,* L. M. J. V. van Straaten (ed.), 54–61. Amsterdam: Elsevier.

Callahan, J. E. 1980. Heavy minerals in stream sediments from Churchill Falls, Labrador – an aid in bedrock mapping. Can. J. Earth Sci. **17,** 244–53.

Callahan, J. 1987. A nontoxic heavy liquid and inexpensive filters for separation of mineral grains. *J. Sedim. Petrol.* **57,** 765–6.

Callender, D. L. & R. L. Folk 1958. Idiomorphic zircon, key to volcanism in the Lower Tertiary sands of Texas. *Am. J. Sci.* **256,** 257–69.

Carroll, D. 1940. Possibilities of heavy mineral correlation of some Permian sedimentary rocks, New South Wales. *Bull. Am. Ass. Petrol. Geol.* **24,** 636–48.

Carroll, D. 1953. Weatherability of zircon. *J. Sedim. Petrol.* **23,** 106–16.

Carver, R. E. 1971. Heavy mineral separation. In *Procedures in sedimentary petrology,* R. E. Carver (ed.), 427–52. New York: Wiley.

Cawood, P. A. 1983. Modal composition of detrital clinopyroxene geochemistry of lithic sandstones from the New England Fold Belt (east Australia): a Paleozoic forearc terrain. *Bull. Geol. Soc. Amer.* **94,** 1199–214.

Chatterjee, B. K. 1966. New technique for preparing polished thin sections of heavy mineral residue. *J. Sedim. Petrol.* **36,** 268–9.

Cheel, R. J. 1984. Heavy mineral shadows, a new sedimentary structure formed under upper-flow-regime conditions: its directional and hydraulic significance. *J. Sedim. Petrol.* **54,** 1175–82.

Chinner, G. A., J. V. Smith & C. R. Knowles 1969. Transition-metal contents of Al_2SiO_5 polymorphs. *Am. J. Sci.* **267-A,** 96–113.

Coleman, R. G. & R. C. Erd 1961. Hydrozircon from the Wind River Formation, Wyoming, USA. *US Geol Survey Prof. Pap.* 424-C, 297–300.

Colman, S. M. & D. P. Dethier (eds) 1986. *Rates of chemical weathering of rocks and minerals.* Orlando: Academic Press.

Dana, E. S. 1895. *The system of mineralogy of James Dwight Dana,* 6th edn. New York: Wiley.

Darby, D. A. & Y. W. Tsang 1987. Variation in ilmenite element composition within and among drainage basins: implications for provenance. *J. Sedim. Petrol.* **57,** 831–8.

Davis, D. K. & W. R. Moore 1970. Dispersal of Mississippi sediment in the Gulf of Mexico: *J. Sedim. Petrol.* **40**, 339–53.

Davis, J. D. 1980. *Statistics and data analysis in geology.* New York: Wiley.

Deer, W. A., R. A. Howie & J. Zussman 1962. *Rock-forming minerals.* Vol. 5: *Non-silicates.* London: Longman.

Deer, W. A., R. A. Howie & J. Zussman 1963. *Rock-forming minerals.* Vol. 2: *Chain silicates.* London: Longman.

Deer, W. A., R. A. Howie & J. Zussman 1978. *Rock-forming minerals.* 2nd edn. Vol. 2A: *Single-chain silicates.* London: Longman.

Deer, W. A., R. A. Howie & J. Zussman 1982. *Rock-forming minerals.* Vol. 1A: *Orthosilicates.* London: Longman.

Deer, W. A., R. A. Howie & J. Zussman 1986. *Rock-forming minerals,* 2nd edn. Vol. 1B: *Disilicates and ring silicates.* London: Longman.

Demina, M. E. 1970. Transport directions for clastic material of Aptian–Cenomanian deposits in western Turkmenia. *J. Math. Geol.* **2,** 349–69.

Doeglas, O. J. 1940. *The importance of heavy mineral analysis for regional sedimentary petrology.* Rept. Comm. Sedimentation 1939–40. Nat. Res. Council, 102–21.

Drummond, S. E. & S. H. Stow 1979. Hydraulic differentiation of heavy minerals, offshore Alabama and Mississippi. *Bull. Geol. Soc. Amer.* **90,** 1429–57.

Dryden, A. L. 1931. Accuracy in percentage representation of heavy mineral frequencies. *Proc. Ac. Nat. Sci.* **17,** 233–8.

Dryden, A. L. & C. Dryden 1946. Comparative rates of weathering of some common heavy minerals. *J. Sedim. Petrol.* **16,** 91–6.

Dudley, R. J. 1976. The use of cathodoluminescence in the identification of soil minerals. *J. Soil. Sci.* **27,** 487–94.

Edelman, C. H. 1933. *Petrologische provincies in het Nederlandsche Kwartair.* PhD thesis, University of Amsterdam. Amsterdam: D. B. Centen's Uitgevers Maatschappij.

Edelman, C. H. & D. J. Doeglas 1932. Reliktstrukturen detritischer Pyroxene und Amphibole. *Tschermaks Min. Petrol. Mitt.* **42,** 482–90.

Edelman, C. H. & D. J. Doeglas 1934. Über Umwandlungserscheinungen an detritischem Staurolith und anderen Mineralien. *Tschermaks Min. Petrol. Mitt.* **45,** 225–34.

Evans, M. & M. A. Mange-Rajetzky 1991. The provenance of sediments in the Barrême thrust-top basin, Haute-Provence, France. In *Developments in Sedimentary Provenance Studies*, A.C. Morton *et al.* (eds), 323–342. Geol Soc. London Spec. Publ. No. 57.

Faure, M. G. 1977. *Principles of isotope geology.* New York: Wiley.

Feo-Codecido, G. 1956. Heavy mineral techniques and their application to Venezuelan stratigraphy. *Bull. Am. Ass. Petrol. Geol.* **40,** 984–1000.

Fessenden, F. W. 1959. Removal of heavy liquid separates from glass centrifuge tubes. *J. Sedim. Petrol.* **29,** 621.

Fleet, W. F. 1926. Petrological notes on the Old Red Sandstones of the West Midlands. *Geol Mag.* **63,** 505–16.

Fleet, W. F. & F. Smithson 1928. On the occurrence of dark apatite in some British rocks. *Geol Mag.* **65,** 6–8.

Flinter, B. H. 1959. A magnetic separation of some alluvial minerals in Malaya. *Am. Miner.* **44,** 738–51.

Flores, R. M. & G. L. Shideler 1978. Factors controlling heavy mineral variations on the south Texas outer continental shelf, Gulf of Mexico. *J. Sedim. Petrol.* **48,** 269–80.

Flores, R. M. & G. L. Shideler 1982. Discriminant analyses of heavy minerals in beach and dune sediments of the Outer Banks barrier, North Carolina. *Bull. Geol Soc. Amer.* **93,** 409–13.

Force, E. R. 1980. The provenance of rutile. *J. Sedim Petrol.* **50,** 485–8.

Fortey, N. J. & U. McL. Michie 1978. Aegirine of possible authigenic origin in Middle Devonian sediments in Caithness, Scotland. *Miner. Mag.* **42,** 439–42.

Friis, H. 1974. Weathered heavy mineral associations from the young-Tertiary deposits of Jutland, Denmark. *Sedim. Geol.* **12,** 199–213.

Friis, H. 1976. Weathering of a Neogene fluviatile fining upwards sequence at Voervadsbro, Denmark. *Bull. Geol Soc. Den.* **25,** 99–105.

Friis, H., O. B. Nielsen, E. M. Friis & B. E. Balme 1980. Sedimentological and palaeontological investigations of a Miocene sequence at Lavsjberg, central Jutland, Denmark. *Den. Geol. Unders. Arbog* **1979,** 51–67.

Füchtbauer, H. 1964. Sedimentpetrographische Untersuchungen in der älteren Molasse nördlich der Alpen. *Eclogae geol. Helv.* **57,** 157–298.

Füchtbauer, H. 1967. Die Sandsteine in der Molasse nördlich der Alpen. *Geol. Rundsch.* **56,** 266–300.

Füchtbauer, H. 1974. *Sediments and sedimentary rocks,* Vol. 1. Stuttgart: Schweizerbart'sche.

Galehouse, J. S. 1967. Provenance and paleocurrents of the Paso Robles Formation, California. *Bull. Geol Soc. Amer.* **78,** 951–78.

Galehouse, J. S. 1969. Counting grain mounts: number percentage vs. number frequency. *J. Sedim. Petrol.* **39,** 812–15.

Galehouse, J. S. 1971. Point counting. In *Procedures in sedimentary petrology,* R. E. Carver (ed.), 385–407. New York: Wiley.

Galloway, M. C. 1972. Statistical analyses of regional heavy mineral variation, Hawkesbury Sandstone and Narrabeen Group (Triassic), Sidney Basin. *J. Geol Soc. Australia* **19,** 65–76.

Gandolfi, G., L. Paganelli & G. G. Zuffa 1983. Petrology and dispersal pattern in the Marnoso–Arenacea Formation (Miocene, northern Apennines). *J. Sedim. Petrol.* **53,** 493–507.

Gardner, J. V., W. E. Dean & T. L. Vallier 1980. Sedimentology and geochemistry of surface sediments, outer continental shelf, southern Bering Sea. *Marine Geol.* **35,** 299–329.

Gastil, R. G., M. de Lisle & J. Morgan 1967. Some effects of progressive metamorphism on zircons. *Bull. Geol Soc. Amer.* **78,** 879–906.

Gautier, A. M. & E. von Pechmann 1984. Mineral separation by centrifugation with heavy liquids – improvement of a method. *Schweiz. Miner. Petrogr. Mitt.* **64,** 458–64.

Gazzi, P. & G. G. Zuffa 1970. Le arenarie paleogeniche dell' Appennino emiliano. *Miner. Petrogr. Acta* **16,** 97–137.

Gazzi, P., G. G. Zuffa, G. G. Gandolfi & L. Paganelli 1973. Provenienza e dispersione litoranea delle sabbie delle spiagge adriatiche fra le foci dell' Isonzo e del Foglia: inquadramento regionale. *Mem. della Soc. Geol. Italiana* **12,** 1–37.

Gleadow, A. J. W., I. R. Duddy & J. F. Lovering 1983. Fission-track analysis: a new tool for the evaluation of thermal histories and hydrocarbon potential. *J. Australian Petr. Expl. Ass.* **23,** 93–102.
Goldfarb, R. J. 1981. A comparison of different geochemical sampling media and the detection of mineralized skarns, southern Sierra Nevada, California (Abstr.). *Geol Soc. Amer. 30th Ann. Meeting, Abstr. with Progr.* 460.
Goldich, S. S. 1938. A study in rock weathering. *J. Geol.* **46,** 17–58.
Goldstein, A. Jr 1942. Sedimentary petrologic provinces of the northern Gulf of Mexico. *J. Sedim. Petrol.* **12,** 77–84.
Gorbatschev, R. 1962. Secondary sphene in the Mälar sandstone. *Geol. Fören. Stockholm, Förh.* **84,** 32–7.
Görz, H., R. J. R. S. B. Bhalla & E. W. White 1970. *Detailed cathodoluminescence characterization of common silicates.* Pennsylvania State Univ. Spec. Publication no. 70–101, 62–70.
Gravenor, C. P. 1979. The nature of the late Paleozoic glaciation in Gondwana as determined from an analysis of garnets and other heavy minerals. *Can. J. Earth Sci.* **16,** 1137–53.
Gravenor, C. P. & V. A. Gostin 1979. Mechanism to explain the loss of heavy minerals from the Upper Palaeozoic tillites of South Africa and Australia and the late Precambrian tillites of Australia. *Sedimentology* **26,** 707–17.
Gravenor, C. P. & R. K. Leavitt 1981. Experimental formation and significance of etch patterns on detrital garnets. *Can. J. Earth Sci.* **18,** 765–75.
Grimm, W. D. 1973. Stepwise heavy mineral weathering in the Residual Quartz Gravel, Bavarian Molasse (Germany). *Contrib. Sedimentology,* **1,** 103–25.
Guinier, A. 1964. *Théorie et technique de la radiocristallographie.* Paris: Dunod.
Gwyn, Q. H. J. & A. Dreimanis 1979. Heavy mineral assemblages in tillites and their use in distinguishing glacial lobes in the Great Lakes region. *Can. J. Earth Sci.* **16,** 2219–35.

Hails, J. R. 1976. Placer deposits. In *Handbook of strata-bound and stratiform ore deposits.* Vol. 3: *Supergene and surficial ore deposits; textures and fabrics,* K. H. Wolf (ed.), 213–44. Amsterdam: Elsevier.
Hall, M. R. & P. H. Ribbe 1971. An electron microprobe study of luminescence centers in cassiterite. *Am. Miner.* **56,** 31–45.
Hand, B. M. 1967. Differentiation of beach and dune sands using settling velocities of light and heavy minerals. *J. Sedim. Petrol.* **37,** 514–20.
Hansley, P. L. 1986. Relationship of detrital nonopaque heavy minerals to diagenesis and provenance of the Morrison Formation, southwestern San Huan Basin, New Mexico. In *A basin analysis case study; the Morrison Formation, Grants uranium region, New Mexico,* Ch. E. Turner-Peterson, E. S. Santos & N. S. Fishman (eds), 257–76. Am. Ass. Petrol. Geol. Studies in Geology, no. 22.
Hansley, P. L. 1987. Petrologic and experimental evidence for the etching of garnets by organic acids in the Upper Jurassic Morrison Formation, northwestern New Mexico. *J. Sedim. Petrol.* **57,** 666–81.
Harrison, W. E. 1973. Heavy minerals of Horn Island, Northern Gulf of Mexico. *J. Sedim. Petrol.* **43,** 391–5.
Hawthorne, F. C. 1981. *Crystal chemistry of the amphiboles.* In *Amphiboles and other hydrous pyriboles – Mineralogy,* D. R. Veblen (ed.) 1–102. Washington DC: Miner. Soc. Amer.
Hearn, B. C. Jr & E. S. McGee 1983. Garnets in Montana diatremes; a key to prospecting for kimberlites. *Bull. US Geol Surv.* **1604,** 1–33.
Heinrich, E. W. 1965. *Microscopic identification of minerals.* New York: McGraw–Hill.
Hemingway, J. E. & M. Y. Tamar-Agha 1975. The effects of diagenesis on some heavy minerals from the sandstone of the Middle Limestone Group in Northumberland. *Proc. Yorks. Geol Soc.* **40,** 537–46.
Henningsen, D. 1967. Crushing of sedimentary rock samples and its effect on shape and number of heavy minerals. *Sedimentology* **8,** 253–5.
Henry, D. J. & Ch. V. Guidotti 1985. Tourmaline as a petrogenetic indicator mineral: an example from the staurolite grade metapelites of NW Maine. *Am. Miner.* **70,** 1–15.
Hering, O. H. & W. Zimmerle 1963. Simple method of distinguishing zircon, monazite and xenotime. *J. Sedim. Petrol.* **33,** 472–3.
Herzog, L. F., D. J. Marshall & R. F. Babione 1970. *The luminoscope—a view for studying the electron stimulated luminescence for terrestrial, extraterrestrial and synthetic materials under the microscope.* Pennsylvania State Univ. Spec. Publ. no. 70–101, 79–98.
Hess, H. H. 1956. *Notes on the operation of Frantz isodynamic magnetic separator.* pamphlet, Princeton Univ.
Hilmy, M. E., E. Slansky & F. I. Khalaf 1971. Opaque minerals in recent beach sediments of Kuwait, Arabia. *N. Jb. Geol. Paläont. Mh.* **6,** 340–4.
Huang, P. M. & M. Schnitzer (eds) 1986. *Interaction of soil minerals with natural organics and microbes.* Madison: Soil Sci. Soc. Amer. Inc.
Huang, W. H. & W. D. Keller 1970. Dissolution of rock-forming silicate minerals in organic acids: simulated first-stage weathering of fresh mineral surfaces. *Am. Miner.* **55,** 2076–94.
Huang, W. H. & W. D. Keller 1972. Organic acids as agents of chemical weathering of silicate minerals. *Nature* **239,** 149–51.
Hubert, J. F. 1962. A zircon–tourmaline–rutile maturity index and the interdependence of the composition of heavy mineral assemblages with the gross composition and texture of sandstones. *J. Sedim. Petrol.* **32,** 440–50.
Hubert, J. F. 1971. Analysis of heavy mineral assemblages. In *Procedures in sedimentary petrology,* R. E. Carver (ed.), 453–78. New York: Wiley.
Hubert, J. F. & W. J. Neal 1967. Mineral composition and dispersal patterns of deep-sea sands in the western North Atlantic petrologic province. *Bull. Geol Soc. Amer.* **78,** 749–72.
Humbert, R. P. & C. E. Marshall 1943. Mineralogical and chemical studies of soil formation from acid and basic igneous rocks from Missouri. *College of Agriculture, Univ. Missouri Res. Bull.* **359,** 1–60.
Hunter, R. E. 1967. A rapid method for determining weight percentages of universal heavy minerals. *J. Sedim. Petrol.* **37,** 521–9.
Hurst, A. & A. C. Morton 1988. An application of heavy mineral analysis to lithostratigraphy and reservoir modelling in the Oseberg Field, Northern North Sea. *Marine and Petroleum Geology* **5,** 157–69.
Hutton, C. O. 1947. Determination of xenotime. *Am. Miner.* **32,** 141–5.
Hutton, C. O. 1950. Studies of heavy detrital minerals. *Bull. Geol Soc. Amer.* **61,** 635–710.

IJlst, L. 1973. New diluents in heavy liquid mineral separation and an improved method for the recovery of the liquids from the washings. *Am. Miner.* **58,** 1084–7.
Illing, V. C. 1916. The oilfields of Trinidad. *Proc. Geol. Ass.* **27,** 115.
Imbrie, J. & Tj. H. van Andel 1964. Vector analysis of heavy mineral data. *Bull. Geol Soc. Amer.* **75,** 1131–56.
Ingram, R. L. 1971. Sieve analysis. In *Procedures in sedimentary petrology,* R. E. Carver (ed.), 49–67. New York: Wiley.

Juvignè, E. & S. Shipley 1983. Distribution of heavy minerals in the downwind tephra lobe of the May 18, 1980 eruption of the Mount

St. Helens (Washington, USA). *Eiszeitalter u. Gegenwart* **33,** 1–7.

Kelling, G., H. Sheng & D. J. Stanley 1975. Mineralogic composition of sand-sized sediment on the outer margin off the Mid-Atlantic States: assessment of the influence of the ancestral Hudson and other fluvial systems. *Bull. Geol Soc. Amer.* **86,** 853–62.

Khalaf, F. I., A. Al-Kadi & S. Al-Saleh 1985. Mineralogical composition and potential sources of dust fallout deposits in Kuwait, northern Arabian Gulf. *Sedim. Geol.* **42,** 255–78.

Klovan, J. E. & A. T. Miesch 1976. Extended CABFAC and QMODEL computer programs for Q-mode factor analysis of compositional data. *Computers and Geosciences* **1,** 161–78.

Klug, H. P. & L. E. Alexander 1974. *X-ray diffraction procedures.* New York: Wiley.

Knebel, H. J. & J. S. Creager 1974. Heavy minerals of the East-Central Bering Sea. *J. Sedim. Petrol.* **44,** 553–61.

Koen, G. M. 1955. Heavy minerals as an aid to the correlation of sediments of the Karroo system in the northern part of the Union of South Africa. *Trans Geol Soc. S. Africa* **58,** 281–366.

Komar, P. D. 1987. Selective grain entrainment by a current from a bed of mixed sizes: a reanalysis. *J. Sedim. Petrol.* **57,** 203–11.

Komar, P. D. & K. E. Clemens 1986. The relationship between grain's settling velocity and threshold motion under unidirectional currents. *J. Sedim. Petrol.* **56,** 258–66.

Komar, P. D. & B. Cui 1984. The analysis of grain-size measurements by sieving and settling-tube techniques. *J. Sedim. Petrol.* **54,** 603–14.

Komar, P. D. & C. Wang 1984. Processes of selective grain transport and the formation of placers on beaches. *J. Geol.* **92,** 637–55.

Kresten, P., P. Fels & G. Berggren 1975. Kimberlitic zircons – A possible aid in prospecting for kimberlites. *Miner. Deposita.* **10,** 47–56.

Krinsley, D. H. & J. C. Doornkamp 1973. *Atlas of quartz sand surface textures.* Cambridge: Cambridge University Press.

Krumbein, W. C. & F. J. Pettijohn 1938. *Manual of sedimentary petrography.* New York: Appleton–Century.

Krynine, P. D. 1946. The tourmaline group in sediments. *J. Geol.* **54,** 65–87.

Leake, B. E. 1978. Nomenclature of amphiboles. *Miner. Mag.* **42,** 533–63.

Leighton, V. L. & M. E. McCallum 1979. *Rapid evaluation of heavy minerals in stream sediments of the Prairie Divide area of northern Colorado—a tool for kimberlite exploration.* U.S. Geol Surv. Open File Report, 79–761.

Leith, C. J. 1950. Removal of iron oxide coatings from mineral grains. *J. Sedim. Petrol.* **20,** 174–6.

Lemcke, K., W. von Engelhardt & H. Füchtbauer 1953. Geologische und sedimentpetrographische Untersuchungen im Westteil der ungefalteten Molasse des süddeutschen Alpenvorlandes. *Beih. Geol. Jb.* **11**.

LeRibault, L. 1977. *L'exoscopie des quartz.* Paris: Masson.

Leu, R. F. & M. M. Druckman 1982. Preparation of grain mounts. *J. Sedim. Petrol.* **52,** 667.

Lin, I. J., V. Rohrlich & A. Slatkine 1974. Surface microtextures of heavy minerals from the Mediterranean coast of Israel. *J. Sedim. Petrol.* **44,** 1281–95.

Lirong, Ch., F. Shouzhi & M. Yanping 1984. The statistical analysis of the heavy mineral assemblage in the sediments of the East China Sea. *Studia Marina Sinica* **21,** 291–6.

Long, J. V. P. & S. O. Agrell 1965. The cathodoluminescence of minerals in thin section. *Miner. Mag.* **34,** 318–26.

Lowright, R., E. G. Williams & F. Dachille 1972. An analysis of factors controlling deviations in hydraulic equivalence in some modern sands. *J. Sedim. Petrol.* **42,** 635–45.

Ludwig, G. G. 1955. Neue Ergebnisse der Schwermineral- und Kornanalyse im Oberkarbon und Rotliegenden des südlichen und östlichen Harzvorlandes. *Beih. Zeitschr. Geol.* **14**.

Ludwig, G. 1968. Zur Lithologie des 'Kulms' bei Erbendorf/ Oberpfalz (Bayern). *N. Jb. Geol. Paläont. Mh.* 407–12.

Luepke, G. 1980. Opaque minerals as aids in distinguishing between source and sorting effects on beach-sand mineralogy in southwestern Oregon. *J. Sedim. Petrol.* **50,** 489–96.

Luepke, G. (ed.) 1984. *Stability of heavy minerals in sediments.* New York: Van Nostrand Reinhold.

Luepke, G. (ed.) 1985. *Economic analysis of heavy minerals in sediments.* New York: Van Nostrand Reinhold.

McCrone, W. C. & J. G. Delly 1973. *The particle atlas,* 2nd edn. Vol. 2: *The light microscopy atlas.* Michigan: Ann Arbor Sci. Publ.

Mackie, W. 1896. The sands and sandstones of eastern Moray. *Trans Geol Soc. Edinb.* **7,** 148–72.

Mackie, W. 1923a. The principles that regulate the distribution of particles of heavy minerals in sedimentary rocks as illustrated by the sandstones of the north-east of Scotland. *Trans Geol Soc. Edinb.* **11,** 138–64.

Mackie, W. 1923b. The source of the purple zircons in the sedimentary rocks of Scotland. *Trans Geol Soc. Edinb.* **11,** 200–13.

McMaster, R. L. 1960. Mineralogy as an indicator of beach sand movement along the Rhode Island shore. *J. Sedim. Petrol.* **30,** 404–13.

McMaster, R. L. & L. E. Garrison 1966. Mineralogy and origin of southern New England shelf sediments. *J. Sedim. Petrol.* **36,** 1131–42.

McQuivey, R. S. & T. N Keefer 1969. *The relation of turbulence to deposition of magnetite over ripples.* Prof. Pap. US Geol Surv., no. 650-D, 244–47.

Mallik, T. K. 1986. Micromorphology of some placer minerals from Kerala beach, India. *Marine Geol.* **71,** 371–81.

Mange-Rajetzky, M. A. 1979. *The mineralogy and provenance of the Quaternary sediments of the southern Turkish coast between Karatas and Antalya.* Unpubl. DIC thesis, Imperial College, London.

Mange-Rajetzky, M. A. 1981. Detrital blue sodic amphibole in Recent sediments, southern coast, Turkey. *J. Geol Soc. Lond.* **138,** 83–92.

Mange-Rajetzky, M. A. 1983. Sediment dispersal from source to shelf on an active continental margin, S. Turkey. *Marine Geol.* **52,** 1–26.

Mange-Rajetzky, M. A. 1989. The use of heavy mineral analyses in assisting zonation, correlation and provenance studies of clastic reservoirs (Abstract). *Marine and Petroleum Geology,* in press.

Mange-Rajetzky, M. A. & R. Oberhänsli 1982. Detrital lawsonite and blue sodic amphibole in the Molasse of Savoy, France and their significance in assessing Alpine evolution. *Schweiz. Miner. Petrogr. Mitt.* **62,** 415–36.

Mange-Rajetzky, M. A. & R. Oberhänsli 1986. Detrital pumpellyite in the Peri-Alpine Molasse. *J. Sedim. Petrol.* **56,** 112–22.

Mariano, A. N. 1977. The use of cathodoluminescence in evaluation of heavy minerals in beach sands. *Nuclide Spectra* **10,** no. 2.

Mariano, A. N. & P. J. Ring 1975. Europium activated cathodoluminescence in minerals. *Geochim. Cosmochim. Acta* **39,** 649–60.

Marshall, D. J. 1988. *Cathodoluminescence of geological materials.* Boston: Unwin Hyman.

Martens, J. H. C. 1932. Piperin as an immersion medium in sedimentary petrography. *Am. Miner.* **17,** 198–9.

Matter, A. & K. Ramseyer 1985. Cathodoluminescence microscopy as a tool for provenance studies of sandstones. In *Provenance of arenites.* G. G. Zuffa (ed.), 191–211. Dordrecht: Reidel.

Maurer, H. 1982. Oberflächentexturen an Schwermineralkörnern aus der Unteren Süsswassermolasse (Chattien) der Westschweiz. *Eclogae geol. Helv.* **75,** 23–31.

Maurer, H. 1983. Sedimentpetrographische Analysen an Molasseabfolgen der Westschweiz. *Jb. Geol. Bundesanstalt* **126,** 23–69.

Maurer, H. & W. Nabholz 1980. Sedimentpetrographie in der Molasse-Abfolge der Bohrung Romanens-1 und in der benachtbarten subalpinen Molasse (Kt. Fribourg). *Eclogae geol. Helv.* **71,** 205–22.

Maurer, H., H. P. Funk & W. Nabholz 1978. Sedimentpetrographische Untersuchungen an Molasseabfogen der Bohrung Linden-1 und ihrer Umgebung (Kt. Bern). *Eclogae geol. Helv.* **71,** 497–516.

Mertie, J. B. Jr 1979. *Monazite in the granitic rocks of the southeastern Atlantic States—an example of the use of heavy minerals in geologic exploration.* Prof. Pap. US Geol Surv., no. 1094.

Middleton, L. T. & M. J. Kraus 1980. Simple technique for thin section preparation of unconsolidated materials. *J. Sedim. Petrol.* **50,** 622–3.

Milner, H. B. 1929. *Sedimentary petrography.* London: Thomas Murby.

Milner, H. B. 1962. *Sedimentary petrography,* 4th edn. London: George Allen & Unwin.

Milton, Ch. & H. P. Eugster 1959. Mineral assemblages of the Green River Formation. In *Researches in geochemistry,* P. H. Abelson (ed.), 118–50. New York: Wiley.

Milton, Ch., B. Ingram & I. Breger 1974. Authigenic magnesioarfvedsonite from the Green River Formation, Duchesne County, Utah. *Am. Miner.* **59,** 830–6.

Morad, S. & A. A. Aldahan 1985. Leucoxene–calcite–quartz aggregates in sandstones and the relation to decomposition of sphene. *N. Jb. Miner. Mh.* **10,** 458–68.

Morton, A. C. 1979a. Surface features of heavy mineral grains from Palaeocene sands of the central North Sea. *Scott. J. Geol.* **15,** 293–300.

Morton, A. C. 1979b. The provenance and distribution of the Palaeocene sands of the central North Sea. *J. Petroleum Geol.* **2,** 11–21.

Morton, A. C. 1982a. The provenance and diagenesis of Palaeogene sandstones of southeast England as indicated by heavy mineral analysis. *Proc. Geol Ass.* **93,** 263–74.

Morton, A. C. 1982b. Lower Tertiary sand development in Viking Graben, North Sea. *Bull. Am. Ass. Petrol. Geol.* **66,** 1542–59.

Morton, A. C. 1983. The mineralogy of Palaeogene sediments in southeast England. Reply. *Proc. Geol. Ass.* **94,** 274–8.

Morton, A. C. 1984a. Heavy minerals from the Palaeogene sediments. Deep Sea Drilling Project Leg 81: their bearing on stratigraphy, sediment provenance and the evolution of the North Atlantic. In *Initial Reports of the Deep Sea Drilling Project,* D. G. Roberts & D. Schnitker *et al.* (eds), Vol. 81, 653–61.

Morton, A. C. 1984b. Stability of detrital heavy minerals in Tertiary sandstones from the North Sea Basin. *Clay Minerals* **19,** 287–308.

Morton, A. C. 1985a. Heavy minerals in provenance studies. In *Provenance of arenites,* G. G. Zuffa (ed.), 249–77. Dordrecht: Reidel.

Morton, A. C. 1985b. A new approach to provenance studies: electron microprobe analysis of detrital garnets from Middle Jurassic sandstones of the North Sea. *Sedimentology* **32,** 553–66.

Morton, A. C. 1986. Dissolution of apatite in North Sea Jurassic sandstones: implications for the generation of secondary porosity. *Clay Minerals* **21,** 711–33.

Morton, A. C. 1987. Influences of provenance and diagenesis on detrital garnet suites in the Paleocene Forties sandstone, central North Sea. *J. Sedim. Petrol.* **57,** 1027–32.

Muir, I. D. 1977. Microscopy: transmitted light. In *Physical methods in determinative mineralogy,* 2nd edn. J. Zussman (ed.), 35–108. London: Academic Press.

Muller, L. D. 1977. Laboratory methods of mineral separation. In *Physical methods in determinative mineralogy,* 2nd edn, J. Zussman (ed.), 1–34. London: Academic Press.

Nahon, D. B. & F. Colin 1982. Chemical weathering of orthopyroxenes under lateritic conditions. *Am. J. Sci.* **282,** 1232–43.

Nayudu, Y. 1962. Rapid method for studying silt-size sediments and heavy minerals by liquid immersion. *J. Sedim. Petrol.* **32,** 326–33.

Nesse, W. D. 1986. *Introduction to optical mineralogy.* New York: Oxford University Press.

Nickel, E. 1973. Experimental dissolution of light and heavy minerals in comparison with weathering and intrastratal solution. *Contrib. Sedimentology* **1,** 1–68.

Nickel, E. 1978. The present status of cathodoluminescence as a tool in sedimentology. *Minerals Science Engineering* **10,** 73–100.

Norman, T. N. 1969. A method to study the distribution of heavy mineral abundance in a turbidite. *Sedimentology* **13,** 263–80.

Norrish, K. & B. W. Chappell 1977. X-ray fluorescence spectrometry. In *Physical methods in determinative mineralogy,* 2nd edn, J. Zussman (ed.), 201–72. London: Academic Press.

Odin, G. S. & A. Matter 1981. De glauconiarum origine. *Sedimentology* **28,** 611–41.

Owen, M. R. 1987. Hafnium content of detrital zircons, a new tool for provenance study. *J. Sedim. Petrol.* **57,** 824–30.

Parfenoff, A., C. Pomerol & J. Tourenq 1970. *Les minéroux en grains.* Méthodes d'étude et détermination. Paris: Masson.

Peterson, C. D., P. D. Komar & K. F. Schneidegger 1986. Distribution, geometry, and origin of heavy mineral placer deposits on Oregon beaches. *J. Sedim. Petrol.* **56,** 67–77.

Pettijohn, F. J. 1941. Persistence of heavy minerals and geologic age. *J. Geol.* **49,** 610–25.

Pettijohn, F. J., P. E. Potter & R. Siever 1973. *Sand and sandstone.* New York: Springer Verlag.

Phillips, W. R. & D. T. Griffen 1981. *Optical mineralogy: the nonopaque minerals.* San Francisco: Freeman, Cooper.

Pichler, H. & C. Schmitt-Riegraf 1987. *Gesteinsbildende Minerale im Dünnschliff.* Stuttgart: Ferdinand Enke.

Piller, H. 1951. Über den Schwermineralgehalt von anstehendem und verwittertem Brockengranit nördlich St. Andreasberg. *Heidelb. Beitr. Min. Petrogr.* **2,** 523–37.

Pirkle, E. C., F. L. Pirkle, W. A. Pirkle & P. R. Stayert 1984. The Yulee heavy mineral sand deposits of northeastern Florida. *Econ. Geol.* **79,** 725–37.

Pirkle, F. L., E. E. Pirkle, A. Pirkle & S. E. Dichs 1985. Evaluation through correlation and principal component analyses of a delta origin for the Hawthorne and Citronelle sediments of peninsular Florida. *J. Geol.* **93,** 493–501.

Poldervaart, A. 1955. Zircon in rocks, 1: sedimentary rocks. *Am. J. Sci.* **253,** 433–61.

Poldervaart, A. 1956. Zircon in rocks, 2: igneous rocks. *Am. J. Sci.* **254,** 521–54.

Poldervaart, A. & H. H. Hess 1951. Pyroxenes in the crystallization of basaltic magma. *J. Geol.* **59,** 472–89.

Portnov, A. M. & B. S. Gorobets 1969. Luminescence of apatite from different rock types. *Dokl. Akad. Nauk. SSSR.* **184,** 110–14.

Power, G. M. 1968. Chemical variation in tourmalines from south west England. *Miner. Mag.* **36,** 1078–89.

Pryor, W. A. & N. C. Hester 1969. X-ray diffraction analysis of heavy minerals. *J. Sedim. Petrol.* **39,** 1384–9.

Pupin, J. P. 1976. *Signification des caractères morphologiques du zircon commun des roches en pétrologie. Base de la méthode typologique. – Applications.* Unpubl. PhD thesis, Univ. Nice.

Pupin, J. P. & G. Turco 1972. Une typologie originale du zircon accessoire. *Bull. Soc. fr. Minéral. Cristallogr.* **95,** 348–59.

Pupin, J. P. & G. Turco 1981. Le zircon, mineral commun significatif des roches endogènes et exogènes. *Bull. Minéral.* **104,** 724–31.

Raeburn, C. & H. B. Milner 1927. *Alluvial prospecting.* London: Thomas Murby.

Raeside, J. D. 1959. Stability of index minerals in soils with particular reference to quartz, zircon and garnet. *J. Sedim. Petrol.* **29,** 493–502.

Ramseyer, K. 1983. *Bau eines Kathodenlumineszenz-Mikroskopes und Diagenese-Untersuchungen an permischen Sedimenten aus Oman.* Unpubl. PhD thesis, Univ. Bern.

Reed, R. D. & Bailey, J. P. 1927. Surface correlation by means of heavy minerals. *Bull. Am. Ass. Petrol Geol.* **11,** 359–68.

Reid, A. M., T. E. Bunch, A. J. Kohen & S. S. Pollack 1964. Luminescence of orthopyroxenes. *Nature* **204,** 1292–3.

Retgers, J. W. 1895. Über die mineralogische und chemische Zusammensetzung der Dünensande Hollands. *N. Jb. Miner.* **1,** 16–74.

Rice, R. M., D. S. Gorsline & R. H. Osborne 1976. Relationships between sand input from rivers and the composition of sands from the beach of southern California. *Sedimentology* **23,** 689–703.

Riech, V., H. R. Kudrass & M. Wiedicke 1982. Heavy minerals of the eastern Australian shelf sediments between Newcastle and Fraser Island. *Geol. Jb.* **56,** 179–95.

Rittenhouse, G. 1943. The transportation and deposition of heavy minerals. *Bull. Geol Soc. Amer.* **54,** 1725–80.

Rittenhouse, G. & W. E. Bertholf Jr 1942. Gravity versus centrifuge separation of heavy minerals from sands. *J. Sedim. Petrol.* **12,** 85–9.

Rizzini, A. 1974. Holocene sedimentary cycle and heavy mineral distribution, Romagna Marche Coastal Plain, Italy. *Sedim. Geol.* **11,** 17–37.

Robson, D. A. 1984. Scanning electron micrographs of some common heavy minerals from the sandstones of the Northumberland Trough. *Trans Nat. History Soc. Northumbria* **52,** 27–34.

Robson, S. H. 1982. *Using the scanning electron microscope and energy dispersive X-ray spectrometer to do mineral identification and compositional point counting on unconsolidated marine sediments.* Techn. Rep. Marine Geosci. Unit Geol. Surv. Cape Town, no. 13, 19–27.

Rosenblum, S. 1958. Magnetic susceptibilities of minerals in the Frantz Isodynamic magnetic separator. *Am. Miner.* **43,** 170–3.

Rubey, W. W. 1933. The size-distribution of heavy minerals within a water-laid sandstone. *J. Sedim. Petrol.* **3,** 3–29.

Russel, R. D. 1937. Mineral composition of Mississippi river sands. *Bull. Geol Soc. Amer.* **48,** 1307–48.

Sallenger, A. H. Jr 1979. Inverse grading and hydraulic equivalence in grain flow deposits. *J. Sedim. Petrol.* **49,** 553–62.

Sanders, J. E. & J. H. Kravitz 1964. Mounting and polishing mineral grains on a microscopic slide for study with reflected and polarized light. *Econ. Geol.* **59,** 291–8.

Sarkisyan, S. G. 1958. Upper Permian continental Molasses of the Pre-Urals. *Eclogae geol. Helv.* **51,** 1043–51.

Scavnicar, R. 1979. Pjescenjaci Pliocena i Miocena sauske potoline. *Zbornik Radova, sekcija za primljenu geologiju, geofiziku, geokemiju, Serija A* **6,** 351–82.

Schluger, P. R. 1976. Petrology and origin of the red beds of the Perry Formation, New Brunswick, Canada and Maine, USA. *J. Sedim. Petrol.* **46,** 22–37.

Schnitzer, W. A. 1983. Zur Problematik der Schwermineralanalyse am Beispiel triassischer Sedimentgesteine. *Geol. Rundsch.* **72,** 67–75.

Schott, J., R. A. Berner & E. L. Sjöberg 1981. Mechanism of pyroxene and amphibole weathering, I: experimental studies of iron-free minerals. *Geochim. Cosmochim. Acta.* **45,** 2123–35.

Scull, B. J. 1960. Removal of heavy liquid separates from glass centrifuge tubes – alternate method. *J. Sedim. Petrol.* **30,** 626.

Setlow, L. W. 1978. Age determination of reddened coastal dunes in northwest Florida, USA, by use of scanning electron microscopy. In *Scanning electron microscopy in the study of sediments,* W. B. Whalley (ed.), 283–305. Norwich: Geo Books.

Setlow, L. W. & R. P. Karpovich 1972. 'Glacial' microtextures on quartz and heavy mineral sand grains from the littoral environment. *J. Sedim. Petrol.* **42,** 864–75.

Shukri, N. M. 1949. The mineralogy of some Nile sediments. *Q. J. Geol Soc. Lond.* **105,** 511–29.

Silver, L. T. & I. D. Williams 1981. Zircon and isotopes in sedimentary basin analyses; a case study of the upper Morrison Formation, southern Colorado Plateau (Abstr.) *Geol Soc. Amer. 94th Ann. Meeting, Abstr. with Programs* **13,** 554–5.

Silver, L. T. & I. D. Williams 1982. Application of zircons and their isotope systems in fluvial sedimentology analysis (Abstr.). *Geol Soc. Amer. Cordilleran Section 78th Ann. Meeting, Abstr. with Programs* **14,** 234.

Simpson, G. S. 1976. Evidence of overgrowths on, and dissolution of, detrital garnets. *J. Sedim. Petrol.* **46,** 689–93.

Sindowsky, K. H. 1938. Sedimentpetrographische Methoden zur Untersuchung sandiger Sedimente. *Geol. Rundsch.* **29,** 196–200.

Sindowsky, K. H. 1958. Schüttungsrichtungen und Mineralprovinzen im westdeutschen Buntsantstein. *Geol. Jb.* **73,** 277–94.

Slingerland, R. L. 1977. The effect of entrainment on the hydraulic equivalence relationships of light and heavy minerals in sands. *J. Sedim. Petrol.* **47,** 753–70.

Slingerland, R. L. 1984. The role of hydraulic sorting in the origin of fluvial placers. *J. Sedim. Petrol.* **54,** 137–50.

Smith, J. V. & R. C. Stenstrom 1965. Electron-excited luminescence as a petrologic tool. *J. Geol.* **73,** 627–35.

Smithson, F. 1928. Geological studies in the Dublin District, I: the heavy minerals of the granite and contiguous rocks in the Ballycorus District. *Geol Mag.* **65,** 12–25.

Smithson, F. 1959. A simple spectroscopic eye-piece for testing monazite under the microscope. *Min. Mag.* **32,** 176.

Sommerauer, J. 1976. *Die chemisch–physikalische Stabilität natürlicher Zirkone und ihr U–(Th)–Pb System.* Unpubl. PhD thesis, Swiss Federal Inst. of Technology, Zürich.

Spears, D. A. 1982. The recognition of volcanic clays and the significance of heavy minerals. *Clay Minerals* **17,** 373–5.

Speer, J. A. 1980. Zircon. In *Reviews in mineralogy,* P. H. Ribbe (ed.), Vol. 5: *Orthosilicates,* 67–112. Washington DC: Miner. Soc. Amer.

Spencer, C. W. 1960. Method for mounting silt-size heavy minerals for identification by liquid immersion. *J. Sedim. Petrol.* **30,** 498–500.

Staatz, M. H., K. J. Murata & J. J. Glass, 1955. Variation of composition and physical properties of tourmaline with its position in the pegmatite. *Am. Miner.* **40,** 789–804.

Stapor, F. W. Jr 1973. Heavy mineral concentrating processes and density/shape/size equilibra in the marine and coastal dune sands of the Apalachicola, Florida, region. *J. Sedim. Petrol.* **43,** 396–407.

Stattegger, K. 1976. Schwermineraluntersuchungen in den klastischen Serien der Variszischen Geosynklinale der Ost- und Zentral-pyrenäen. *Mitt. österr. geol. Ges.* **69**, 267–90.

Stattegger, K. 1982. Schwermineraluntersuchungen in der östlichen Grauwackenzone (Steiermark/Oesterreich) und deren statistische Auswertung. *Verh. Geol. Bundesanstalt* **2,** 107–22.

Stattegger, K. 1987. Heavy minerals and provenance of sands: modelling of lithological end members from river sands of northern

Austria and from sandstones of the austroalpine Gosau Formation (late Cretaceous). *J. Sedim. Petrol.* **57,** 301–10.

Steidtman, J. R. 1982. Size-density sorting of sand-size spheres during deposition from bedload transport and implications concerning hydraulic equivalence. *Sedimentology* **29,** 877–83.

Stewart, R. A. 1986. Routine heavy mineral analysis using a concentrating table. *J. Sedim. Petrol.* **56,** 555–6.

Stieglitz, R. D. 1969. Surface textures of quartz and heavy mineral grains from fresh-water environments: an application of scanning electron microscopy. *Bull. Geol. Soc. Amer.* **80,** 2091–4.

Stumpfl, E. 1958. Erzmikroskopische Untersuchungen an Schwermineralien in Sanden. *Geol. Jb.* **73,** 685–723.

Swift, D. J. P., Ch. E. Dill & J. McHone 1971. Hydraulic fractionation of heavy mineral suites on an unconsolidated retreating coast. *J. Sedim. Petrol.* **41,** 683–90.

Tan, K. H. 1986. Degradation of soil minerals by organic acids. In *Interactions of soil minerals with natural organics and microbes,* P. H. Huang & M. Schnitzer (eds), 1–25. Madison: Soil. Sci. Amer. Inc.

Thomas, R. B., A. E. Burford & L. L. Chy 1984. Characteristics of rare earth elements in the heavy minerals of the sediments in San Luis Valley, Colorado. *Geol Soc. Amer. 97th Ann. Meeting, Abstr. with Programs* **16,** 675.

Thoulet, J. 1881. Etude minéralogique d'un sable du Sahara. *Bull. Soc. Minér. France* **4,** 262–8.

Thoulet, J. 1913. Notes de lithologie sous-marine. *Ann. Inst. Oceanogr.* **5,** 1–14.

Tickell, F. G. 1965. *The techniques of sedimentary mineralogy.* Amsterdam: Elsevier.

Tieh, Th. T. 1973. Heavy mineral assemblages in some Tertiary sediments in San Mateo County, California. *J. Sedim. Petrol.* **43,** 408–17.

Tomita, T. 1954. Geologic significance of the colour of granite zircon and the discovery of the Precambrian in Japan. *Kyushu Univ. Mem. Fac. Sci. Ser. D. Geol.* **4,** 135–61.

Trask, C. B. & B. M. Hand 1985. Differential transport of fall-equivalent sand grains, Lake Ontario, New York. *J. Sedim. Petrol.* **55,** 226–34.

Tröger, W. E. 1969. *Optische Bestimmung der gesteinsbildenden Minerale.* Vol. 2: 2nd edn. Stuttgart: Schweizerbart'sche.

van Andel, Tj. H. 1950. *Provenance, transport and deposition of Rhine sediments.* Wageningen: H. Veeman en Zonen.

van Andel, Tj. H. 1955. Recent sediments of the Rhone delta II. Sources and deposition of heavy minerals. *Geol. Mijnb.* **15,** 515–56.

van Andel, Tj. H. 1959. Reflection on the interpretation of heavy mineral analyses. *J. Sedim. Petrol.* **29,** 153–63.

van Andel, Tj. H. 1960. Sources and dispersion of Holocene sediments, northern Gulf of Mexico. In *Recent sediments, Northwest Gulf of Mexico,* F. P. Shepard, F. B. Phleger & Tj. H. van Andel (eds), 33–55. Tulsa: Publ. Am. Ass. Petrol. Geol.

van Andel, Tj. H. 1964. Recent marine sediments of Gulf of California. In *Marine geology of the Gulf of California,* Tj. H. van Andel & G. G. Shor Jr (eds), 216–310. Mem. Am. Ass. Petrol. Geol., no. 3.

van Andel, Tj. H. & D. M. Poole 1960. Sources of Recent sediments in the northern Gulf of Mexico. *J. Sedim. Petrol.* **30,** 91–122.

van Baren, F. A. & H. Kiel 1950. Contribution to the sedimentary petrology of the Sunda Shelf. *J. Sedim. Petrol.* **20,** 185–213.

van Harten, D. 1965. On the estimation of relative grain frequencies in heavy mineral slides. *Geol. Mijnb.* **44,** 357–63.

van Hilten, D. 1981. Refractive indices of minerals through the microscope: a simple method by oblique observation. *Am. Miner.* **66,** 1089–91.

Vatan, A. 1949. La sédimentation détritique dans la zone subalpine et la Jura méridional au Crétacé et au Tertiaire. *C. R. somm. Séance Soc. géol. France* **6,** 102–4.

Velbel, M. A. 1984. Natural weathering mechanism of almandine garnet. *Geology* **12,** 631–4.

von Erffa, A. 1973. Sedimentation, Transport und Erosion an der Nordküste Kolumbiens zwischen Barranquilla und der Sierra Nevada de Santa Marta. *Mitt. Ins. Colombo-Aleman Invest. Cient.* **7,** 155–209.

Walker, T. R. 1967. Formation of red-beds in modern and ancient deserts. *Bull. Geol Soc. Amer.* **78,** 353–68.

Walker, T. R., P. H. Ribbe & R. M. Honea 1967. Geochemistry of hornblende alteration in Pliocene red beds, Baja California, Mexico. *Bull. Geol Soc. Amer.* **78,** 1055–60.

Wang, Ch. & P. D. Komar 1985. The sieving of heavy mineral sands. *J. Sedim. Petrol.* **55,** 479–82.

Weaver, C. E. 1963. Interpretative value of heavy minerals from bentonites. *J. Sedim. Petrol.* **33,** 343–9.

Weiblen, P. 1965. Investigation of cathodoluminescence with the petrographic microscope. In *Developments in applied spectroscopy,* Vol. 4, E. N. Davis (ed.) 245–51. New York: Plenum.

Weissbrod, T. & J. Nachmias 1986. Stratigraphic significance of heavy minerals in the late Precambrian–Mesozoic clastic sequence ('Nubian Sandstone') in the Near East. *Sedim. Geol.* **47,** 263–91.

Weyl, R. & H. Werner 1951. *Schwermineraluntersuchungen im Jungtertiär und Altquartär Schleswig-Holsteins.* Proc. 3rd. Internat. Sedim. Congr. Groningen–Wageningen, 293–303.

Wieseneder, H. 1952. Die Verteilung der Schwermineralien im nördlichen Inneralpinen Wiener Becken und ihre geologische Deutung. *Verhandl. Geol. Bundesanstalt,* 207–22.

Weiseneder, H. & J. Maurer 1958. Ursachen der räumlichen und zeitlichen Aenderung des Mineralbestandes der Sedimente des Wiener Beckens. *Ecologae geol. Helv.* **51,** 1155–72.

Wilson, M. J. 1975. Chemical weathering of some primary rock-forming minerals. *Soil Sci.* **119,** 349–55.

Winkler, W. & D. Bernoulli 1986. Detrital high-pressure/low-temperature minerals in late Turonian flysch sequence of the eastern Alps (western Austria): implications for early Alpine tectonics. *Geology* **14,** 598–601.

Winkler, W., G. Galetti & M. Maggetti 1985. Bentonite im Gurnigel-, Schlieren- und Wägital-Flysch: Mineralogie, Chemismus, Herkunft. *Eclogae geol. Helv.* **78,** 545–64.

Woletz, G. 1963. Chrakteristische Abfolgen der Schwermineralgehalte in Kreide- und Alttertiär-Schichten der nördlichen Ostalpen. *Jb. geol. Bundesanstalt* **106,** 89–119.

Woletz, G. 1967. Schwermineralvergesellschaftungen aus ostalpinen Sedimentationsbecken der Kreidezeit. *Geol. Rundsch.* **56,** 308–20.

Yurkova, R. M. 1970. Comparison of postsedimentary alteration of oil–gas and waterbearing rocks. *Sedimentology* **15,** 53–68.

Zeschke, G. 1961. Prospecting for ore deposits by panning heavy minerals from river sands. *Econ. Geol.* **56,** 1250–7.

Zimmerle, W. 1972. Sind detritsche Zirkone rötlicher Farbe auch in Mitteleuropa Indikatoren für präkambrische Liefergebiete? *Geol. Rundsch.* **61,** 116–39.

Zimmerle, W. 1984. The geotectonic significance of detrital brown spinel in sediments. *Mitt. Geol.-Paläont Inst. Univ. Hamburg* **56,** 337–60.

Zinkernagel, U. 1978. Cathodoluminescence of quartz and its application to sandstone petrology. *Contrib. Sedimentology* **8.**

Zussman, J. 1977. X-ray diffraction. In *Physical methods in determinative mineralogy,* 2nd edn., J. Zussman (ed.), 391–473. London: Academic Press.

Identification table

MINERALS		Colourless	Pink	Orange	Red	Purple, violet	Blue	Green	Yellow	Brown	Gray	Black
Isotropic	Garnet	*	*	o	o			o		o		
	Spinel-Group				*		*	*		*		
	Sphalerite	o							*	*		
	Fluorite	*	o				o					
Uniaxial (+)	Zircon	*	*		o	o	o		o		o	o
	Cassiterite	*	o		*				*	*		o
	Rutile				*	o			*	*		o
	Xenotime	*						*	*	o		
	Scheelite	*							o	o	o	
Uniaxial (-)	Vesuvianite	*					o	o	o			
	Tourmaline	o	o				o	*	o	*		o
	Corundum	*	o				*					
	Anatase	o					*		*	*		
	Jarosite								*	o		
	Calcite	*	o									
	Siderite	*							*	*		
	Dolomite-Ankerite	*							*		o	
	Apatite	*						o		o	o	
Biaxial (+)	Sphene	*						o	*	o		
	Sillimanite, Fibrolite	*							o		o	
	Topaz	*	o		o		o					
	Staurolite								*	*		
	Zoisite	*	o									
	Clinozoisite	*						o	o			
	Piemontite		*			o			o			
	Lawsonite	*					o					
	Pumpellyite	o						*	o			
	Enstatite	*						o				
	Diopside	*						o				
	Diallage	*						o	o			
	Augite	o				o		*	o	*		
	Prehnite	*										
	Brookite			o					*	*		
	Baryte	*					o				o	
	Celestite	*					o					
	Gypsum	*										
	Anhydrite	*	o	o								
	Monazite	o						*	*	o		
Biaxial (+) (-)	Olivine	*						o	o			
	Chloritoid						*				o	
	Allanite									*		
	Aegirine-Augite							o				
	Anthophyllite-Gedrite	*							o	o		
	Hornblende-Series							*		*		
	Riebeckite						*	o		o		
	Chlorite-Group							o	o			
Biaxial (-)	Vesuvianite	*					o	o	o			
	Andalusite	*	*									
	Kyanite	*					o					
	Dumortierite	*				*	*					
	Epidote							*	o			
	Axinite	o				o	o		*	*		
	Hypersthene		*		*			*				
	Aegirine (Acmite)							*		*		
	Tremolite	*						o			o	
	Actinolite							*				
	Glaucophane-Crossite					o	*					
	Arfvedsonite						*	o		o		
	White Mica	*									o	
	Glauconite							*	o	o		
	Biotite							o	o	*		
	Talc	*										
	Serpentine Group	*						*	*	o	o	

* Common colour o Infrequent colour

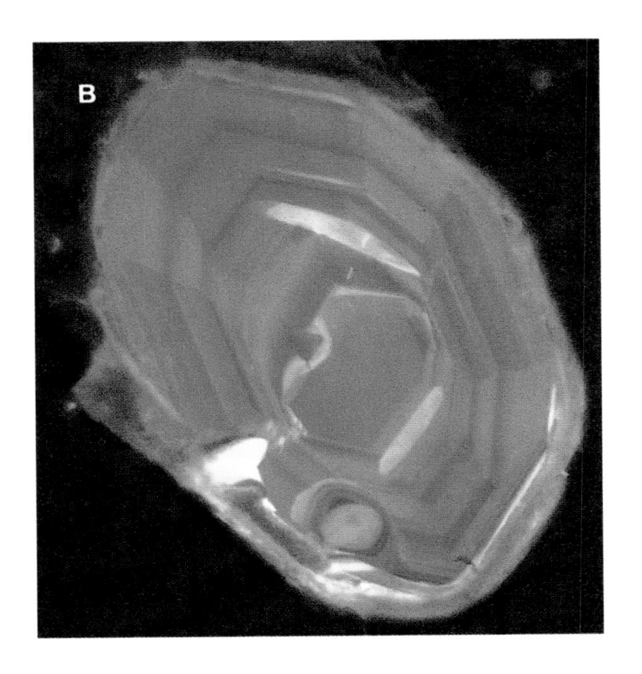

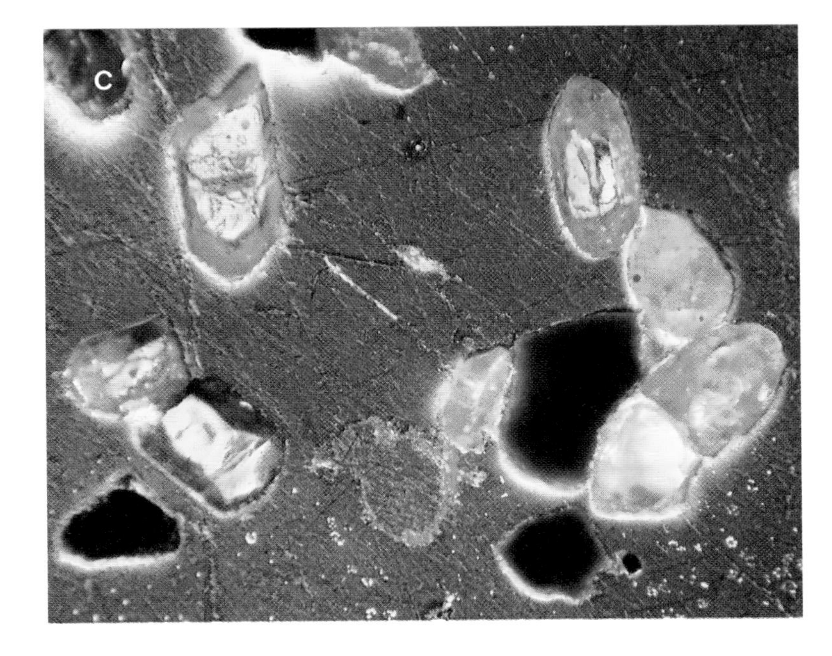

Figure 3.5 Cathodoluminescence photomicrographs
(A) Zircons extracted from an orthogneiss. Grains show blue luminescence and internal zoning. Note the curved terminations and rounded morphology of many grains (Maggia Massif, Switzerland). (B) Purple zircon (hyacinth) exhibiting blue luminescence and distinct internal zoning (Diamond placer, West Africa). (C) Zircons with yellow luminescing cores and non-luminescing overgrowth (Cretaceous, northern Tunisia).

Mineral Index

General Index

References in italics are to figures